AF680671

Solid State Chemistry

Solid State Chemistry

Bernadett Kent

STATES
ACADEMIC PRESS
www.statesacademicpress.com

Published by States Academic Press
109 South 5th Street,
Brooklyn, NY 11249, USA
www.statesacademicpress.com

Solid State Chemistry
Bernadett Kent

© 2022 States Academic Press

International Standard Book Number: 978-1-63989-490-1 (Hardback)

This book contains information obtained from authentic and highly regarded sources. All chapters are published with permission under the Creative Commons Attribution Share Alike License or equivalent. A wide variety of references are listed. Permissions and sources are indicated; for detailed attributions, please refer to the permissions page. Reasonable efforts have been made to publish reliable data and information, but the authors, editors and publisher cannot assume any responsibility for the validity of all materials or the consequences of their use.

Trademark Notice: Registered trademark of products or corporate names are used only for explanation and identification without intent to infringe.

Cataloging-in-Publication Data

Solid state chemistry / Bernadett Kent.
p. cm.
Includes bibliographical references and index.
ISBN 978-1-63989-490-1
1. Solid state chemistry. 2. Chemistry, Physical and theoretical. 3. Chemistry. I. Kent, Bernadett.
QD478 .S65 2022
541.042 1--dc23

Table of Contents

Preface

This book aims to help a broader range of students by exploring a wide variety of significant topics related to this discipline. It will help students in achieving a higher level of understanding of the subject and excel in their respective fields. This book would not have been possible without the unwavered support of my senior professors who took out the time to provide me feedback and help me with the process. I would also like to thank my family for their patience and support.

Solid state chemistry refers to the study of structure, properties and synthesis of solid materials. It focuses on the making and characterization of the constituent particles of novel materials. Solids are divided into two categories based on the arrangement of their particles and the nature of order present. These categories are crystalline and amorphous solids. A wide variety of techniques are used to study these properties of solids. Some of these are oven techniques like gas reaction, melt methods and solution methods along with chemie douce. The field primarily overlaps with other areas such as thermodynamics, crystallography, solid state physics, ceramics, material science and electronics. This book is a valuable compilation of topics, ranging from the basic to the most complex theories and principles related to this discipline. It studies, analyzes and upholds the pillars of solid state chemistry and its utmost significance in modern times. For someone with an interest and eye for detail, this book covers the most significant topics of solid state chemistry.

A brief overview of the book contents is provided below:

Chapter - What is Solid State Chemistry?

The branch of chemistry that deals with the study of the synthesis, structure, and properties of solid phase materials is referred to as solid state chemistry. Crystal structure, crystal systems, crystal symmetry, Miller indices, etc. are studied under its domain. This is an introductory chapter which will briefly introduce about solid state chemistry.

Chapter - Classification of Solids

Solids can be basically classified into two categories namely crystalline and amorphous solids. Crystalline solids are well ordered at the atomic level, and amorphous solids are disordered. It includes ceramics, composite materials, polymers, etc. The topics elaborated in this chapter will help in gaining a better perspective of different classifications of solids.

Chapter - Bonding in Solids

Solids can also be classified as per the nature of the bonding between their molecular and atomic components. Four major types of bonding are covalent bonding, ionic bonding, metallic bonding and inter-molecular bonding. This chapter closely examines different types of bonding in solids to provide an extensive understanding of the subject.

Chapter - Solid-State Synthesis

There are many processes and reactions that fall under solid-State synthesis. Some of them are sol-gel process, chemical transport reaction, solid-gas reaction, co-precipitation synthesis, hydrothermal synthesis, intercalation, etc. This chapter sheds light on the subject of solid-State synthesis for an in-depth understanding of it.

Chapter - Crystal Structure and Defects

Crystal defects refer to the imperfections in the regular patterns of the atoms in crystalline solids. Different defects include vacancy defect, Schottky defect, Frenkel defect, interstitial defect, Stone-Wales defect, etc. This chapter has been carefully written to provide an easy understanding of the subject of crystal defects and non-stoichiometry.

Bernadett Kent

1

What is Solid State Chemistry?

The branch of chemistry that deals with the study of the synthesis, structure, and properties of solid phase materials is referred to as solid state chemistry. Crystal structure, crystal systems, crystal symmetry, Miller indices, etc. are studied under its domain. This is an introductory chapter which will briefly introduce about solid state chemistry.

Solid-state chemistry is the study of the synthesis, structure, chemical, and physical properties and the application of solid materials.

A solid is a material in the solid state. Solid state chemistry is the branch of chemistry that deals with the representation of the structure, properties and applications, for example in mineralogy and crystallography, metallurgy and in the Materials Sciences of these substances.

The focus of solid state chemistry will be placed on the consideration of inorganic, crystalline and non-molecular solids, which differ in their reactions, properties and behavior of liquid and gaseous chemical systems.

Properties of Solid

As you should remember from the kinetic molecular theory, the molecules in solids are not moving in the same manner as those in liquids or gases. Solid molecules simply vibrate and rotate in place rather than move about. Solids are generally held together by ionic or strong covalent bonding, and the attractive forces between the atoms, ions, or molecules in solids are very strong. In fact, these forces are so strong that particles in a solid are held in fixed positions and have very little freedom of movement. Solids have definite shapes and definite volumes and are not compressible to any extent.

There are two main categories of solids—crystalline solids and amorphous solids. Crystalline solids are those in which the atoms, ions, or molecules that make up the solid exist in a regular, well-defined arrangement. The smallest repeating pattern of crystalline solids is known as the unit cell, and unit cells are like bricks in a wall—they are all identical and repeating. The other main type of solids are called the amorphous solids. Amorphous solids do not have much order in their structures. Though their molecules are close together and have little freedom to move, they are not arranged in a regular order as are those in crystalline solids. Common examples of this type of solid are glass and plastics.

Physical Properties

Physical properties of elements and compounds that provide conclusive evidence of chemical composition include odor, color, volume, density (mass per unit volume), melting point, boiling point, heat capacity, physical form and shape at room temperature (solid, liquid or gas; cubic, trigonal crystals, etc.), hardness, porosity, index of refraction and many others.

Mechanical

Granite rock formation in the Chilean Patagonia. Like most inorganic minerals formed by oxidation in the Earth's atmosphere, granite consists primarily of crystalline silica SiO_2 and alumina Al_2O_3.

The mechanical properties of materials describe characteristics such as their strength and resistance to deformation. For example, steel beams are used in construction because of their high strength, meaning that they neither break nor bend significantly under the applied load.

Mechanical properties include elasticity and plasticity, tensile strength, compressive strength, shear strength, fracture toughness, ductility (low in brittle materials), and indentation hardness. Solid mechanics is the study of the behavior of solid matter under external actions such as external forces and temperature changes.

A solid does not exhibit macroscopic flow, as fluids do. Any degree of departure from its original shape is called deformation. The proportion of deformation to original size is called strain. If the applied stress is sufficiently low, almost all solid materials behave in such a way that the strain is directly proportional to the stress (Hooke's law). The coefficient of the proportion is called the modulus of elasticity or Young's modulus.

This region of deformation is known as the linearly elastic region. Three models can describe how a solid responds to an applied stress:

- Elasticity: When an applied stress is removed, the material returns to its undeformed state.
- Viscoelasticity: These are materials that behave elastically, but also have damping. When the applied stress is removed, work has to be done against the damping effects and is converted to heat within the material. This results in a hysteresis loop in the stress–strain curve. This implies that the mechanical response has a time-dependence.
- Plasticity: Materials that behave elastically generally do so when the applied stress is less than a yield value. When the stress is greater than the yield stress, the material behaves plastically and does not return to its previous state. That is, irreversible plastic deformation (or viscous flow) occurs after yield that is permanent.

Many materials become weaker at high temperatures. Materials that retain their strength at high temperatures, called refractory materials, are useful for many purposes. For example, glass-ceramics have become extremely useful for countertop cooking, as they exhibit excellent mechanical properties and can sustain repeated and quick temperature changes up to 1000 °C. In the aerospace industry, high performance materials used in the design of aircraft and/or spacecraft exteriors must have a high resistance to thermal shock. Thus, synthetic fibers spun out of organic polymers and polymer/ceramic/metal composite materials and fiber-reinforced polymers are now being designed with this purpose in mind.

Thermal

Because solids have thermal energy, their atoms vibrate about fixed mean positions within the ordered (or disordered) lattice. The spectrum of lattice vibrations in a crystalline or glassy network provides the foundation for the kinetic theory of solids. This motion occurs at the atomic level, and thus cannot be observed or detected without highly specialized equipment, such as that used in spectroscopy.

Thermal properties of solids include thermal conductivity, which is the property of a material that indicates its ability to conduct heat. Solids also have a specific heat capacity, which is the capacity of a material to store energy in the form of heat (or thermal lattice vibrations).

Electrical

Electrical properties include conductivity, resistance, impedance and capacitance. Electrical conductors such as metals and alloys are contrasted with electrical insulators such as glasses and ceramics. Semiconductors behave somewhere in between. Whereas conductivity in metals is caused by electrons, both electrons and holes contribute to current in semiconductors. Alternatively, ions support electric current in ionic conductors.

Many materials also exhibit superconductivity at low temperatures; they include metallic elements such as tin and aluminium, various metallic alloys, some heavily doped semiconductors, and certain ceramics. The electrical resistivity of most electrical (metallic) conductors generally decreases gradually as the temperature is lowered, but remains finite. In a superconductor, however, the resistance drops abruptly to zero when the material is cooled below its critical temperature. An electric current flowing in a loop of superconducting wire can persist indefinitely with no power source.

A dielectric, or electrical insulator, is a substance that is highly resistant to the flow of electric current. A dielectric, such as plastic, tends to concentrate an applied electric field within itself, which property is used in capacitors. A capacitor is an electrical device that can store energy in the electric field between a pair of closely spaced conductors (called 'plates'). When voltage is applied to the capacitor, electric charges of equal magnitude, but opposite polarity, build up on each plate. Capacitors are used in electrical circuits as energy-storage devices, as well as in electronic filters to differentiate between high-frequency and low-frequency signals.

Electro-mechanical

Piezoelectricity is the ability of crystals to generate a voltage in response to an applied mechanical stress. The piezoelectric effect is reversible in that piezoelectric crystals, when subjected to an externally applied voltage, can change shape by a small amount. Polymer materials like rubber, wool, hair, wood fiber, and silk often behave as electrets. For example, the polymer polyvinylidene fluoride (PVDF) exhibits a piezoelectric response several times larger than the traditional piezoelectric material quartz (crystalline SiO_2). The deformation (~0.1%) lends itself to useful technical applications such as high-voltage sources, loudspeakers, lasers, as well as chemical, biological, and acousto-optic sensors and/or transducers.

Optical

Materials can transmit (e.g. glass) or reflect (e.g. metals) visible light.

Many materials will transmit some wavelengths while blocking others. For example, window glass is transparent to visible light, but much less so to most of the frequencies of ultraviolet light that cause sunburn. This property is used for frequency-selective optical filters, which can alter the color of incident light.

For some purposes, both the optical and mechanical properties of a material can be of interest. For example, the sensors on an infrared homing ("heat-seeking") missile must be protected by a cover that is transparent to infrared radiation. The current material of choice for high-speed infrared-guided missile domes is single-crystal sapphire. The optical transmission of sapphire does not actually extend to cover the entire mid-infrared range (3–5 μm), but starts to drop off at wavelengths greater than approximately 4.5 μm at room temperature. While the strength of sapphire is better than that of other available mid-range infrared dome materials at room temperature, it weakens above 600 °C. A long-standing trade-off exists between optical bandpass and mechanical durability; new materials such as transparent ceramics or optical nanocomposites may provide improved performance.

Guided lightwave transmission involves the field of fiber optics and the ability of certain glasses to transmit, simultaneously and with low loss of intensity, a range of frequencies (multi-mode optical waveguides) with little interference between them. Optical waveguides are used as components in integrated optical circuits or as the transmission medium in optical communication systems.

Opto-electronic

A solar cell or photovoltaic cell is a device that converts light energy into electrical energy. Fundamentally, the device needs to fulfill only two functions: photo-generation of charge carriers (electrons and holes) in a light-absorbing material, and separation of the charge carriers to a conductive contact that will transmit the electricity (simply put, carrying electrons off through a metal contact into an external circuit). This conversion is called the photoelectric effect, and the field of research related to solar cells is known as photovoltaics.

Solar cells have many applications. They have long been used in situations where electrical power from the grid is unavailable, such as in remote area power systems, Earth-orbiting satellites and space probes, handheld calculators, wrist watches, remote radiotelephones and water pumping applications. More recently, they are starting to be used in assemblies of solar modules (photovoltaic arrays) connected to the electricity grid through an inverter, that is not to act as a sole supply but as an additional electricity source.

All solar cells require a light absorbing material contained within the cell structure to absorb photons and generate electrons via the photovoltaic effect. The materials used in solar cells tend to have the property of preferentially absorbing the wavelengths of solar light that reach the earth surface. Some solar cells are optimized for light absorption beyond Earth's atmosphere, as well.

Electrical Properties

Electric Conductivity of a substance is defined as its ability to transmit heat energy or electric energy (and in some cases also sound energy). So a good conductor of electricity is easily able to transmit the energy without boiling or melting or changing its composition in any way.

Solids have varying degrees of conductivity i.e. all solids do not have uniform electrical properties. Actually, solids can be divided into three broad categories based on their electrical conductivity. These three groups are as follows:

Conductors

Broadly speaking, conductors are solids that have good electrical conductivity. They allow heat energy and electric currents to transmit through them with ease and speed. Conductors allow this transfer of energy to happen via free flow of electrons from atom to atom. They have the ability to carry this energy throughout themselves even when the current is only applied to one part of their body.

All metals are known to be the best conductors. Their conductivity depends on the number of valence electrons of their atoms. These electrons are not closely bound together and can move freely. Metals have such electrons in their atoms, and so they conduct heat and electricity so well. Metals allow the electric field to transmit through them in conductivities order of $10^6 - 10^8$ ohm^{-1}.

Insulators

Stark opposite to conductors, insulators are materials that absolutely do not conduct any electric energy or currents. They do not let any (or very little) electric charge flow through them. They have a significant band gap which prevents the flow of electricity. Some examples are glass, wood, plastics, rubber etc.

Since insulators are very bad conductors they have another use. We use them to insulate conductors and semiconductors. For example, you must have seen copper wires with a covering of plastic or some form of polymers. They protect the wires and cables without allowing the electric current to go through them. This is insulation of the wire.

Semiconductors

Semiconductors are the go-between conductors and insulators. These are solids that have the ability to conduct electricity through them but only under certain conditions. There are two such conditions that affect the ability of semiconductors to conduct energy, heat, and impurities.

- Intrinsic Semiconductor: These are pure materials with no impurities added, so they are known as undoped semiconductors. Here we apply thermal energy to the material and create vacancies in the valence bands. This allows the energy to pass through. And yet these are not very good conductors and find very limited applications.
- Extrinsic Semiconductors: These are doped semiconductors. We add certain impurities to increase the conductivity of the materials. There are two types of extrinsic semiconductor, namely n-type and p-type. Examples are when we boost the conductivity of silicon and germanium by this method.

One reason why semiconductors are an important material is that we have control over their conductivity. This is why they find many applications, especially in electronics.

Magnetic Properties

The magnetic properties of a solid are the result of the magnetic property of the atoms or ions of these solids. More specifically the magnetism and magnetization of a solid will depend on the movement of electrons in an atom. It can thus be said that each electron of an atom behaves like a magnet, lending the whole solid its magnetic property.

This magnetic behaviour of the electrons of an atom is due to the movement patterns. They have specifically two types of movements:

- Electrons revolve around the nucleus of the atom.
- Electrons also spin on their own axis, spins in opposite sides are labelled with + and – signs.

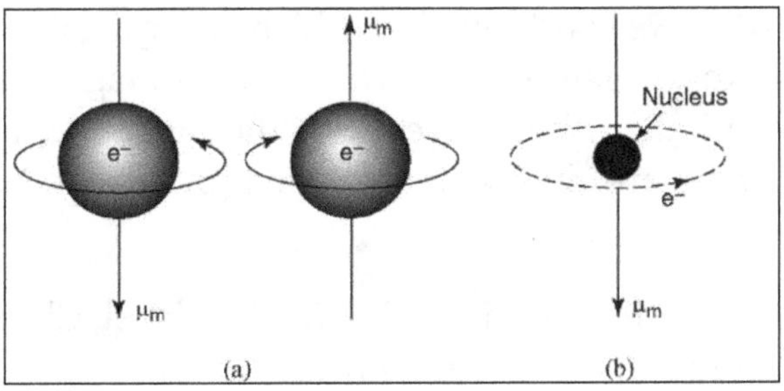

These two motions of the electrons give the atom and the substance their magnetic power. These constant motions make an electric field around the electrons, almost like a loop of current which lends it its magnetic property. On the basis of their magnetic properties, solids can be classified into five categories.

Paramagnetic

These substances are weakly magnetized in an external magnetic field. The direction is the same direction of the magnetic field. So they gain a net magnetization When we remove the paramagnetic substance from the field, the alignment of electrons is interrupted and the substance will lose its magnetic property. Thus paramagnetic substances are not permanent magnets.

Paramagnetism is due to at least one pair of unpaired electrons in its orbit shell which get magnetized in the magnetic field. Some common examples are O_2, Cu_2 etc. These paramagnetic substances find a variety of applications in electronics.

Diamagnetic

Just like paramagnetism, in diamagnetism too the substances are magnetized in an external magnetic field. But diamagnetic solids are repelled in the field. The magnetic property effected in them is in the opposite direction of the magnetic fields and thus they have a repulsive force.

In diamagnetic substances, all electrons in their last shell are paired, there are no valence electrons. This is the reason that the magnetic moment of their atoms is nearly zero. Examples are substances like Sodium Chloride, Benzene etc. Being such bad conductors, we use them as insulators.

Ferromagnetic

Now, these solids are strongly magnetized when we place them in an external magnetic field. Besides the very strong attraction forces, these solids can actually be magnetized permanently. This means that even when the external magnetic fields are removed the solids will retain their magnetic properties.

It is a widely believed theory that the ferromagnetic structures have certain special characteristics. They have what we call "domains' which is a special grouping of metal ions. Each domain is similar to a small magnet. In an electromagnetic field, these domains rearrange themselves and align themselves with the magnetic field. In a non-magnetized metal, these domains are randomly arranged and it cancels out their magnetic properties.

The examples of ferromagnetic solids include Cobalt, Nickel, Chromium Oxide etc. and they have widespread industrial and every day uses.

Antiferromagnetic

In antiferromagnetism, the domain structures of the solid are very similar to those of ferromagnetic solids. But here the domains are oppositely oriented. This means they cancel out each other's magnetism.

Ferrimagnetic

These substances occur when magnetic moments are aligned in both directions (parallel as well as anti-parallel) but in unequal numbers. These are weakly attracted to magnetic fields. Also on heating, these substances will lose their ferrimagnetism altogether. Examples are magnetite and ferrites of Zinc and Magnesium.

References

- Solid-state-chemistry, chemistry: internetchemistry.com, Retrieved 02 March, 2020
- Electrical-properties-of-solids, chemistry-the-solid-state: toppr.com, Retrieved 27 June, 2020
- Crystal-structure, chemistry: byjus.com, Retrieved 17 April, 2020

- Solid-state-braggs-law: thechemistryguru.com, Retrieved 23 August, 2020
- Atomic-structure: byjus.com, Retrieved 14 March, 2020
- Miller-indices, science: britannica.com, Retrieved 26 March, 2020
- Crystal-symmetry, science-dictionaries-thesauruses-pictures-and-press-releases: encyclopedia.com, Retrieved 08 February, 2020

2
Classification of Solids

Solids can be basically classified into two categories namely crystalline and amorphous solids. Crystalline solids are well ordered at the atomic level, and amorphous solids are disordered. It includes ceramics, composite materials, polymers, etc. The topics elaborated in this chapter will help in gaining a better perspective of different classifications of solids.

Crystalline and Amorphous

Crystalline Solids the solids featuring highly ordered arrangements of their particles (atoms, ions, and molecules) in microscopic structures are called crystalline solids.

These ordered microscopic structures make up a crystal lattice that accounts for the structure of the solid at any given point. Examples of crystalline solids include salt (sodium chloride), diamond, and sodium nitrate.

Amorphous Solids

The solids in which the particles are not arranged in any specific order or the solids that lack the overall order of a crystal lattice are called amorphous solids.

The term 'amorphous', when broken down into its Greek roots, can be roughly translated to "without form". Many polymers are amorphous solids. Other examples of such solids include glass, gels, and nanostructured materials.

An ideal crystal is defined as an atomic arrangement that has infinite translational symmetry in all the three dimensions, whereas such a definite definition is not possible for an ideal amorphous solid (a-solid).

Key features of crystalline and amorphous solids.

Nature

- Crystalline Solids: True solids.

- Amorphous Solids: Pseudo-solids or super-cooled liquids.

Geometry

- Crystalline Solids: Particles are arranged in a repeating pattern. They have a regular and ordered arrangement resulting in a definite shape.
- Amorphous Solids: Particles are arranged randomly. They do not have an ordered arrangement resulting in irregular shapes

Melting Points

- Crystalline Solids: They have a sharp melting point.
- Amorphous Solids: They do not have sharp melting points. The solid tends to soften gradually over a temperature range.

Heat of Fusion

(The change in enthalpy when a substance is heated to change its state from solid to liquid.)

- Crystalline Solids: They have definite heat of fusion.
- Amorphous Solids: They do not have definite heat of fusion.

Isotropism

- Crystalline Solids: Anisotropic in nature. i.e., the magnitude of physical properties (such as refractive index, electrical conductivity, thermal conductivity etc) is different along with different directions of the crystal.
- Amorphous Solids: Isotropic in nature. i.e., the magnitude of the physical properties is the same along with all directions of the solid.

Cleavage Property

- Crystalline Solids: When cutting with a sharp edge, the two new halves will have smooth surfaces.
- Amorphous Solids: When cutting with a sharp edge, the two resulting halves will have irregular surfaces.

Rigidity

- Crystalline Solids: They are rigid solids and applying mild forces will not distort its shape.
- Amorphous Solids: They are not rigid, so mild effects may change the shape.

Metals

Traditionally, the term metal (from the Greek word metallon) has been applied to a chemical elementthat has a shiny surface and is a good conductor of heat and electricity. These properties, however, can vary from one metal to the next. More recently, chemists have recognized that the main distinguishing features of a metal are (a) the ability of its atoms to lose some of their outermost electrons to form cations, and (b) the bonding of its atoms by what are called metallic bonds.

Metals form one of three groups of elements—the other two being nonmetals and metalloids. These groups are distinguished by their ionization and bonding properties. On the periodic table, a diagonal line drawn from boron (B) to polonium (Po) separates the metals from the nonmetals. Elements on this line are metalloids, sometimes called semi-metals; elements to the lower left are metals; elements to the upper right are nonmetals. In nature, nonmetals are more abundant than metals, but most elements in the periodic table are metals. Some well-known metals are aluminum, calcium, copper, gold, iron, lead, magnesium, platinum, silver, titanium, uranium, and zinc.

Hot metal work from a blacksmith.

Metals and their alloys are extremely useful for both aesthetic and practical purposes. While some are used for jewelry and works of art, many serve as construction materials for buildings, bridges, transportation vehicles, and industrial machinery. Some, such as copper, are used as conductors in electrical wiring; others, such as platinum and palladium, are catalysts for chemical reactions. Tungsten is used in the filaments of incandescent light bulbs; uranium and plutonium are used in nuclear weapons and nuclear power plants. Moreover, metal ions play significant roles in biological processes, such as the functions of enzymes, the transport of oxygen, and the health of teeth and bones.

Properties

All metals, except mercury, are solids at room temperature. In addition, the colors of metals tend to range from silvery white to gray—the two exceptions are copper and gold. As mentioned above, metals tend to be lustrous (shiny) and good conductors of heat and electricity. They also conduct sound well. Furthermore, they are usually ductile and malleable—that is, they can be readily drawn into wires and beaten into sheets. Solid nonmetals, on the other hand, are generally brittle, lack luster, and are insulators.

Metals are usually thought of as being hard substances, high in density and melting point. It should be noted, however, that there are metals that are soft, low in density, and low in melting point, but they are also quite reactive, and we rarely encounter them in their elemental, metallic form. For example, lithium, sodium, and potassium are less dense than water and are also highly reactive. On the other hand, some of the densest metals are osmium, iridium, platinum, and gold. The melting point of cesium is only 28.4 °C, but that of tungsten is as high as 3,422 °C.

The property of electrical conductivity is mainly because each atom exerts only a loose hold on its outermost electrons, called the valence electrons). Thus, a pure metal may be thought of as a lattice of positively charged ions (cations) surrounded by a cloud of delocalized electrons.

Most metals are chemically unstable, reacting with oxygen in the air to form oxides. Depending on the metal, the time scale of the reaction can vary. The alkali metals (in Group 1 of the periodic table) react quickest, followed by the alkaline earth metals (in group 2). The transition metals—such as iron, copper, zinc, and nickel—take much longer to oxidize. For example, potassium burns in seconds, but iron rusts gradually over a long period of time, depending on the prevailing environmental conditions.

Other metals—such as palladium, platinum, and gold—do not react with the atmosphere at all. Some metals form a barrier layer of oxideon their surface, which cannot be penetrated by further oxygen molecules. They thus retain their shiny appearance and good conductivity for many decades. Examples are aluminum and titanium. The oxides of metals are basic (as opposed to those of nonmetals, which are acidic).

Metal cations combine with nonmetal anions to form salts. Some common classes of salts are carbonates, sulfides, sulfates, silicates, chlorides, nitrates, and phosphates. Many of the minerals found in nature are salts.

Base Metals

In alchemy, the term base metal was used as a designation for common and inexpensive metals, to draw a contrast with precious metals such as gold and silver. A long-cherished goal of the alchemists was the transmutation of base metals into precious metals.

In chemistry today, the term base metal is used informally to refer to a metal that oxidizes or corrodes relatively easily and reacts variably with dilute hydrochloric acid (HCl) to form hydrogen. Examples include iron, nickel, lead, and zinc. Copper, too, is considered a base metal because it oxidizes relatively easily, although it does not react with HCl. Metals that resist oxidation or corrosion are called noble metals, which also tend to be precious metals.

In the past, coins used to derive their value primarily from their precious metal content. Most modern currencies, however, are fiat currency—money that is given legal tender status. This approach allows coins to be made of base metals.

Precious Metals

A gold nugget.

A precious metal is a rare metallic chemical element of high, durable economic value. The best-known precious metals are gold and silver. Although both have industrial uses, they are better known for their uses in art, jewelry, and coinage. Other precious metals include the platinum group metals: ruthenium, rhodium, palladium, osmium, iridium, and platinum, of which platinum is the most widely traded. Plutonium and uranium may also be considered precious metals.

Chemically, the precious metals are less reactive than most elements. They have high luster and higher melting points than other metals. Historically, precious metals were important as currency, but are now regarded mainly as investment and industrial commodities. Investments in gold and silver are often regarded as a hedge against inflation-and economic downturn.

Bullion

A 500-gram silver bullion bar produced by Johnson Matthey.

Precious metals in bulk form are known as bullion and are traded on commodity markets. Bullion metals may be cast into ingots, or minted into coins. The defining attribute of bullion is that it is valued by its mass and purity rather than by a face value as money.

An American Platinum Eagle bullion coin.

The minting of bullion coins by national governments gives them some numismatic value in addition to their bullion value, as well as certifying their purity. The level of purity varies from country to country. Some bullion coins, such as the Canadian Gold Maple Leaf, are available at a purity as high as 99.99%. 100 percent pure bullion is not possible, because the absolute purity of extracted and refined metals can only be approached asymptotically.

One of the largest bullion coins in the world is a 10,000 Australian dollar coin that consists of a full kilogram (kg) of 99.9 percent pure gold. China, however, has produced limited quantities of coins (less than 20 pieces) that exceed 260 troy ounces (8 kg) of gold.

Silver bullion coins have become popular with coin collectors because of their relative affordability. Unlike most gold and platinum issues, which are valued based on prevailing markets, silver issues are more often valued as collectibles—far higher than their actual bullion value.

Precious Metal Status

Gold bars from the Bank of Sweden.

A given metal is precious if it is rare. If mining or refining processes improve, or new supplies are discovered and exploited, the value of such a metal declines.

An interesting case of a precious metal going common is that of aluminum. It is one of the Earth's most common metals, but when first discovered, it was extremely difficult to separate from its ores. For a while, aluminum was regarded as more valuable than gold, and aluminum bars were exhibited alongside the French crown jewels at the Exposition Universelle. Napoleon Bonaparte is said to have used aluminum plates for his most honored guests at dinner. Its price, however, began to drop and collapsed altogether when an easy extraction method, the Hall-Héroult process, was discovered in 1886.

Alloys

An alloy is a combination of two or more chemical elements, at least one of which is a metal, and where the resulting material has metallic properties. Examples of alloys are steel (iron and carbon), brass (copper and zinc), bronze (copper and tin), and duralumin (aluminum and copper). The resulting metallic substance generally has properties significantly different from those of its components.

An alloy with two components is called a binary alloy; one with three is a ternary alloy; one with four is a quaternary alloy. Alloys specially designed for highly demanding applications, such as jet engines, may contain more than ten elements. When an alloy contains mercury, it is called an amalgam.

An alloy "inherits" the characteristics of the elements it was made from, but it is usually designed to have properties more desirable than those of its components. For instance, steel is stronger than iron, one of its main components.

Unlike pure metals, most alloys do not have a sharp melting point. Instead, they have a melting range in which the material becomes a mixture of solid and liquid phases. The temperature at which melting begins is called the solidus, and that at which melting is complete is called the liquidus. For most pairs of elements, however, there is a particular ratio that has a single melting point, and this is called a eutectic mixture.

In practice, some alloys are named after their primary constituent. For example, 14 carat (58 percent) gold is an alloy of gold with other elements. Similarly, the silver used in jewelry and the aluminum used as a structural material are alloys.

Astronomy

In the specialized usage of astronomy and astrophysics, the term "metal" is often used in referring to any element other than hydrogen or helium—it includes substances as chemically nonmetallic as neon, fluorine, and oxygen. Given that the temperatures of stellar bodies allow practically no solid or liquid matter, and little normal chemistry can exist at temperatures that break down almost all chemical bonds, the term "metal" refers to materials produced by a type of nuclear reaction called the triple-helium process.

Uses

Based on their properties, metals have a wide range of uses. For example, based on their strength and hardness, many metals and their alloys are used as structural materials for buildings, bridges, industrial machines, railroad tracks, automobiles, trains, ships, aircraft, spacecraft, and so forth. A number of metals, such as platinum and palladium, are good catalysts for chemical reactions. Precious metals, particularly gold and silver, are used in jewelry and decorative items. Amalgams are used in dental fillings. Copper, an abundant metal, is an extremely good conductor of electricity and is widely used for electrical wiring. Tungsten, with its high melting point, is suitable for use in the filaments of incandescent light bulbs. Uranium and plutonium are radioactive materials used for nuclear weapons and nuclear power plants that generate electricity. In biological systems, metal ions play a number of important roles, such as the functions of enzymes, the transport of oxygen by hemoglobin (which contains iron), and the health of bones and teeth (which need calcium ions).

Crystal Structure of Metals

The Crystalline Nature of Metals

All metallic elements (except Cs, Ga, and Hg) are crystalline solids at room temperature. Like ionic solids, metals and alloys have a very strong tendency to crystallize, whether they are made by thermal processing or by other techniques such as solution reduction or electroplating. Metals crystallize readily and it is difficult to form a glassy metal even with very rapid cooling. Molten metals have low viscosity, and the identical (essentially spherical) atoms can pack into a crystal very easily. Glassy metals can be made, however, by rapidly cooling alloys, particularly if the constituent atoms have different sizes. The different atoms cannot pack in a simple unit cell, sometimes making crystallization slow enough to form a glass.

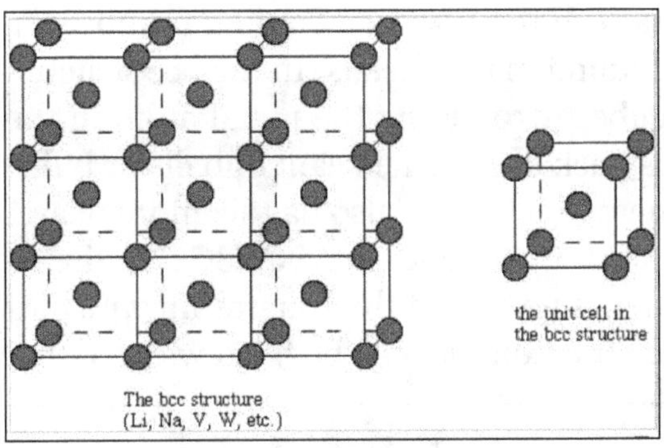

Body-centered cubic.

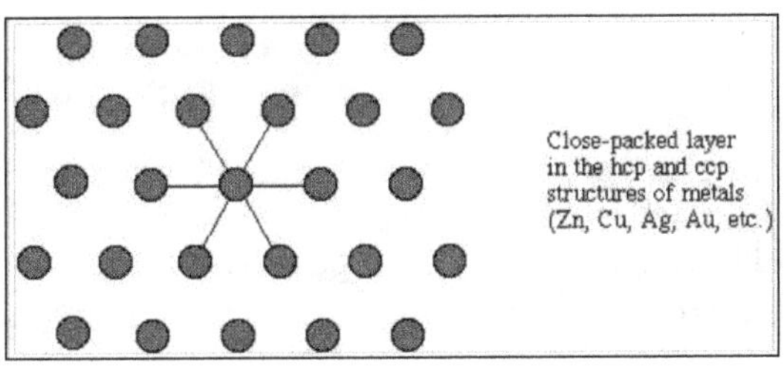

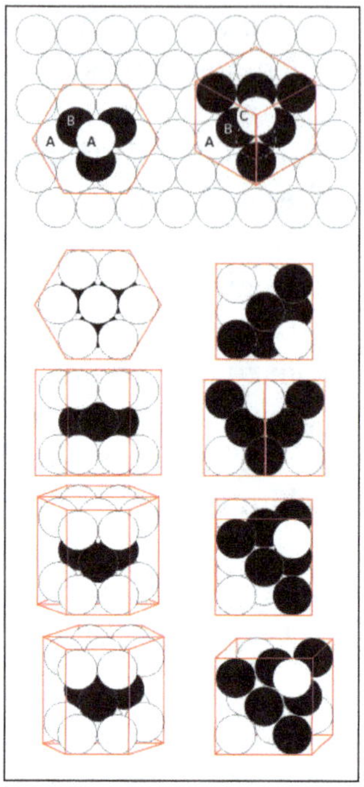

Hcp (left) and fcc (right) close-packing of spheres.

Crystal Structures

Most metals and alloys crystallize in one of three very common structures: body-centered cubic (bcc), hexagonal close packed (hcp), or cubic close packed (ccp, also called face centered cubic, fcc). In all three structures the coordination number of the metal atoms (i.e., the number of equidistant nearest neighbors) is rather high: 8 for bcc, and 12 for hcp and ccp. We can contrast this with the low coordination numbers (i.e., low valences - like 2 for O, 3 for N, or 4 for C) found in nonmetals. In the bcc structure, the nearest neighbors are at the corners of a cube surrounding the metal atom in the center. In the hcp and ccp structures, the atoms pack like stacked cannonballs or billiard balls, in layers with a six-coordinate arrangement. Each atom also has six more nearest neighbors from layers above and below. The stacking sequence is ABCABC... in the ccp lattice and ABAB... in hcp. In both cases, it can be shown that the spheres fill 74% of the volume of the lattice. This is the highest volume fraction that can be filled with a lattice of equal spheres.

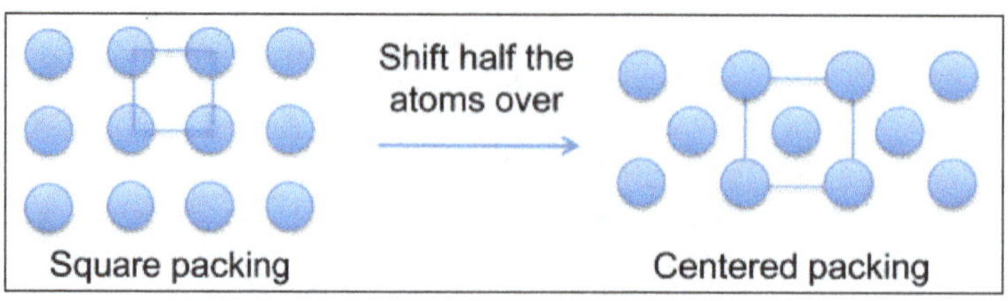

Atoms in metallic crystals have a tendency to pack in dense arrangments that fill space efficiently. The simple square packing (above) upon which the simple cubic structure is based is inefficient and thus rare among metallic crystal structures. Body- or face-centered structures fill space more efficiently and more common.

Periodic Trends in Structure and Metallic Behavior

Remember where we find the metallic elements in the periodic table - everywhere except the upper right corner. This means that as we go down a group in the p-block (let's say, group IVA, the carbon group, or group VA, the nitrogen group), the properties of the elements gradually change from nonmetals to metalloids to metals. The carbon group nicely illustrates the transition. Starting at the top, the element carbon has two stable allotropes - graphite and diamond. In each one, the valence of carbon atoms is exactly satisfied by making four electron pair bonds to neighboring atoms. In graphite, each carbon has three nearest neighbors, and so there are two single bonds and one double bond. In diamond, there are four nearest neighbors situated at the vertices of a tetrahedron, and so there is a single bond to each one.

The two elements right under carbon (silicon and germanium) in the periodic table also have the diamond structure (recall that these elements cannot make double bonds to themselves easily, so there is no graphite allotrope for Si or Ge). While diamond is a good insulator, both silicon and germanium are semiconductors (i.e., metalloids). Mechanically, they are hard like diamond. Like carbon, each atom of Si and Ge satisfies its valence of four by making single bonds to four nearest neighbors.

The next element under germanium is tin (Sn). Tin has two allotropes, one with the diamond structure, and one with a slightly distorted bcc structure. The latter has metallic properties (metallic luster, malleability), and conductivity about 10^9 times higher than Si. Finally, lead (Pb), the element under Sn, has the ccp structure, and also is metallic. Trends in coordination number and conducting properties:

Element	Structure	Coord. No.	Conductivity
C	Graphite, Diamond	3, 4	Semimetal, Insulator
Si	Diamond	4	Semiconductor
Ge	Diamond	4	Semiconductor
Sn	Diamond, Distorted Bcc	4, 8	Semiconductor, Metal
Pb	Ccp	12	Metal

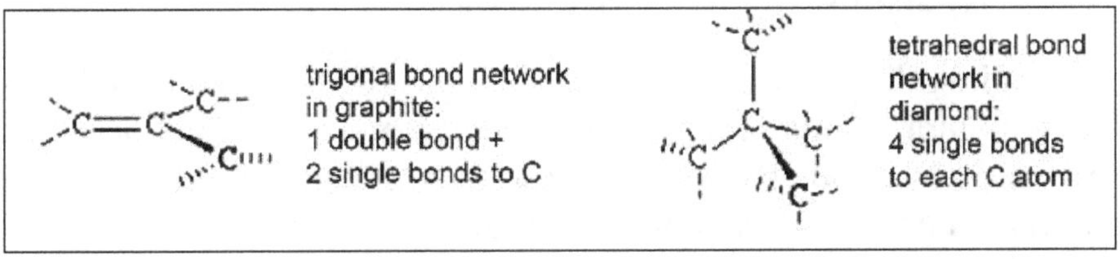

The elements C, Si, and Ge obey the octet rule, and we can easily identify the electron pair bonds in their structures. Sn and Pb, on the other hand, adopt structures with high coordination numbers. They do not have enough valence electrons to make electron pair bonds to each neighbor (this is a common feature of metals). What happens in this case is that the valence electrons become "smeared out" or delocalized over all the atoms in the crystal. It is best to think of the bonding in metals as a crystalline arrangement of positively charged cores with a "sea" of shared valence electrons gluing the structure together. Because the electrons are not localized in any particular bond between atoms, they can move in an electric field, which is why metals conduct electricity well. Another way to describe the bonding in metals is nondirectional. That is, an atom's nearest neighbors surround it in every direction, rather than in a few particular directions (like at the corners of a tetrahedron, as we found for diamond). Nonmetals (insulators and semiconductors), on the other hand, have directional bonding. Because the bonding is non-directional and coordination numbers are high, it is relatively easy to deform the coordination sphere (i.e., break or stretch bonds) than it is in the case of a nonmetal. This is why elements like Pb are much more malleable than C, Si, or Ge.

Minerals

Crystals of serandite, natrolite, analcime, and aegirine from Mont Saint-Hilaire, Quebec, Canada.

A mineral is, broadly speaking, a solid chemical compound that occurs naturally in pure form. Minerals are most commonly associated with rocks due to the presence of minerals within rocks. These rocks may consist of one type of mineral, or may be an aggregate of two or more different types of minerals, spacially segregated into distinct phases. Compounds that occur only in living beings are usually excluded, but some minerals are often biogenic (such as calcite) or are organic compounds in the sense of chemistry (such as mellite). Moreover, living beings often synthesize inorganic minerals (such as hydroxylapatite) that also occur in rocks.

In geology and mineralogy, the term "mineral" is usually reserved for mineral species: crystalline compounds with a fairly well-defined chemical composition and a specific crystal structure. Minerals without a definite crystalline structure, such as opal or obsidian, are then more properly called mineraloids. If a chemical compound may occur naturally with different crystal structures, each structure is considered different mineral species. Thus, for example, quartz and stishovite are two different minerals consisting of the same compound, silicon dioxide.

The International Mineralogical Association (IMA) is the world's premier standard body for the definition and nomenclature of mineral species. As of November 2018, the IMA recognizes 5,413 official mineral species. out of more than 5,500 proposed or traditional ones.

The chemical composition of a named mineral species may vary somewhat by the inclusion of small amounts of impurities. Specific varieties of a species sometimes have conventional or official names of their own. For example, amethyst is a purple variety of the mineral species quartz. Some mineral species can have variable proportions of two or more chemical elements that occupy equivalent positions in the mineral's structure; for example, the formula of mackinawite is given as $(Fe,Ni)_9S_8$, meaning $Fe_xNi_{9-x}S_8$, where x is a variable number between 0 and 9. Sometimes a mineral with variable composition is split into separate species, more or less arbitrarily, forming a mineral group; that is the case of the silicates $Ca_xMg_yFe_{2-x-y}SiO_4$, the olivine group.

Besides the essential chemical composition and crystal structure, the description of a mineral species usually includes its common physical properties such as habit, hardness, lustre, diaphaneity, colour, streak, tenacity, cleavage, fracture, parting, specific gravity, magnetism, fluorescence, radioactivity, as well as its taste or smell and its reaction to acid.

Minerals are classified by key chemical constituents; the two dominant systems are the Dana classification and the Strunz classification. Silicate minerals comprise approximately 90% of the Earth's crust. Other important mineral groups include the native elements, sulfides, oxides, halides, carbonates, sulfates, and phosphates.

Recent Advances

Mineral classification schemes and their definitions are evolving to match recent advances in mineral science. Recent changes have included the addition of an organic class, in both the new Dana and the Strunz classification schemes. The organic class includes a very rare group of minerals with hydrocarbons. The IMA Commission on New Minerals and Mineral Names adopted in 2009 a hierarchical scheme for the naming and classification of mineral groups and group names and established seven commissions and four working groups to review and classify minerals into an official listing of their published names. According to these new rules, "mineral species can be grouped in a number of different ways, on the basis of chemistry, crystal structure, occurrence, association, genetic history, or resource, for example, depending on the purpose to be served by the classification."

The Nickel exclusion of biogenic substances was not universally adhered to. For example, Lowenstam stated that "organisms are capable of forming a diverse array of minerals, some of which cannot be formed inorganically in the biosphere." The distinction is a matter of classification and less to do with the constituents of the minerals themselves. Skinner views all solids as potential minerals and includes biominerals in the mineral kingdom, which are those that are created by the metabolic activities of organisms. Skinner expanded the previous definition of a mineral to classify "element or compound, amorphous or crystalline, formed through *biogeochemical* processes," as a mineral.

Recent advances in high-resolution genetics and X-ray absorption spectroscopy are providing revelations on the biogeochemical relations between microorganisms and minerals that may make Nickel's biogenic mineral exclusion obsolete and Skinner's biogenic mineral inclusion a necessity. For example, the IMA-commissioned "Working Group on Environmental Mineralogy and Geochemistry " deals with minerals in the hydrosphere, atmosphere, and biosphere. The group's scope includes mineral-forming microorganisms, which exist on nearly every rock, soil, and particle surface spanning the globe to depths of at least 1600 metres below the sea floor and 70 kilometres into the stratosphere (possibly entering the mesosphere). Biogeochemical cycles have contributed to the formation of minerals for billions of years. Microorganisms can precipitate metals from solution, contributing to the formation of ore deposits. They can also catalyze the dissolution of minerals.

Prior to the International Mineralogical Association's listing, over 60 biominerals had been discovered, named, and published. These minerals (a sub-set tabulated in Lowenstam) are considered minerals proper according to the Skinner definition. These biominerals are not listed in the International Mineral Association official list of mineral names, however, many of these biomineral representatives are distributed amongst the 78 mineral classes listed in the Dana classification scheme. Another rare class of minerals (primarily biological in origin) include the mineral liquid crystals that have properties of both liquids and crystals. To date, over 80,000 liquid crystalline compounds have been identified.

The Skinner definition of a mineral takes this matter into account by stating that a mineral can be crystalline or amorphous, the latter group including liquid crystals. Although biominerals and liquid mineral crystals, are not the most common form of minerals, they help to define the limits of what constitutes a mineral proper. The formal Nickel definition explicitly mentioned crystallinity as a key to defining a substance as a mineral.

Rocks, Ores and Gems

Schist is a metamorphic rock characterized by an abundance of platy minerals.
In this example, the rock has prominent sillimanite porphyroblasts as large as 3 cm (1.2 in).

Minerals are not equivalent to rocks. A rock is an aggregate of one or more minerals or mineraloids. Some rocks, such as limestone or quartzite, are composed primarily of one mineral – calcite or aragonite in the case of limestone, and quartz in the latter case. Other rocks can be defined by relative abundances of key (essential) minerals; a granite is defined by proportions of quartz, alkali feldspar, and plagioclase feldspar. The other minerals in the rock are termed *accessory minerals*, and do not greatly affect the bulk composition of the rock. Rocks can also be composed entirely of non-mineral material; coal is a sedimentary rock composed primarily of organically derived carbon.

In rocks, some mineral species and groups are much more abundant than others; these are termed the rock-forming minerals. The major examples of these are quartz, the feldspars, the micas, the amphiboles, the pyroxenes, the olivines, and calcite; except for the last one, all of these minerals are silicates. Overall, around 150 minerals are considered particularly important, whether in terms of their abundance or aesthetic value in terms of collecting.

Commercially valuable minerals and rocks are referred to as industrial minerals. For example, muscovite, a white mica, can be used for windows (sometimes referred to as isinglass), as a filler, or as an insulator. Ores are minerals that have a high concentration of a certain element, typically a metal. Examples are cinnabar (HgS), an ore of mercury, sphalerite (ZnS), an ore of zinc, or cassiterite (SnO_2), an ore of tin. Gems are minerals with an ornamental value, and are distinguished from non-gems by their beauty, durability, and usually, rarity. There are about 20 mineral species that qualify as gem minerals, which constitute about 35 of the most common gemstones. Gem minerals are often present in several varieties, and so one mineral can account for several different gemstones; for example, ruby and sapphire are both corundum, Al_2O_3.

Nomenclature and Classification

Minerals are classified by variety, species, series and group, in order of increasing generality. The basic level of definition is that of mineral species, each of which is distinguished from the others by unique chemical and physical properties. For example, quartz is defined by its formula, SiO_2, and a specific crystalline structure that distinguishes it from other minerals with the same chemical formula (termed polymorphs). When there exists a range of composition between two minerals species, a mineral series is defined. For example, the biotite series is represented by variable amounts of the endmembers phlogopite, siderophyllite, annite, and eastonite. In contrast, a mineral group is a grouping of mineral species with some common chemical properties that share a crystal structure. The pyroxene group has a common formula of $XY(Si,Al)_2O_6$, where X and Y are both cations, with X typically bigger than Y; the pyroxenes are single-chain silicates that crystallize in either the orthorhombic or monoclinic crystal systems. Finally, a mineral variety is a specific type of mineral species that differs by some physical characteristic, such as colour or crystal habit. An example is amethyst, which is a purple variety of quartz.

Two common classifications, Dana and Strunz, are used for minerals; both rely on composition, specifically with regards to important chemical groups, and structure. The Dana classification assigns a four-part number to a mineral species. Its class number is based on important compositional groups; the type gives the ratio of cations to anions in the mineral, and the last two numbers group minerals by structural similarity within a given type or class. The less commonly used Strunz classification, named for German mineralogist Karl Hugo Strunz, is based on the Dana system, but combines both chemical and structural criteria, the latter with regards to distribution of chemical bonds.

As of November 2018, 5,413 mineral species are approved by the IMA. They are most commonly named after a person (45%), followed by discovery location (23%); names based on chemical composition (14%) and physical properties (8%) are the two other major groups of mineral name etymologies.

The word "species" comes from the classification scheme in *Systema Naturae* by Carl Linnaeus. He divided the natural world into three kingdoms – plants, animals, and minerals – and classified each with the same hierarchy. In descending order, these were Phylum, Class, Order, Family, Tribe, Genus, and Species.

Chemistry

Hübnerite, the manganese-rich end-member of the wolframite series, with minor quartz in the background.

The abundance and diversity of minerals is controlled directly by their chemistry, in turn dependent on elemental abundances in the Earth. The majority of minerals observed are derived from the Earth's crust. Eight elements account for most of the key components of minerals, due to their abundance in the crust. These eight elements, summing to over 98% of the crust by weight, are, in order of decreasing abundance: oxygen, silicon, aluminium, iron, magnesium, calcium, sodium and potassium. Oxygen and silicon are by far the two most important – oxygen composes 47% of the crust by weight, and silicon accounts for 28%.

The minerals that form are directly controlled by the bulk chemistry of the parent body. For example, a magma rich in iron and magnesium will form mafic minerals, such as olivine and the pyroxenes; in contrast, a more silica-rich magma will crystallize to form minerals that incorporate more SiO_2, such as the feldspars and quartz. In a limestone, calcite or aragonite (both $CaCO_3$) form because the rock is rich in calcium and carbonate. A corollary is that a mineral will not be found in a rock whose bulk chemistry does not resemble the bulk chemistry of a given mineral with the exception of trace minerals. For example, kyanite, Al_2SiO_5 forms from the metamorphism of aluminium-rich shales; it would not likely occur in aluminium-poor rock, such as quartzite.

The chemical composition may vary between end member species of a solid solution series. For example, the plagioclase feldspars comprise a continuous series from sodium-rich end member albite ($NaAlSi_3O_8$) to calcium-rich anorthite ($CaAl_2Si_2O_8$) with four recognized intermediate varieties between them (given in order from sodium- to calcium-rich): oligoclase, andesine, labradorite, and bytownite. Other examples of series include the olivine series of magnesium-rich forsterite and iron-rich fayalite, and the wolframite series of manganese-rich hübnerite and iron-rich ferberite.

Chemical substitution and coordination polyhedra explain this common feature of minerals. In nature, minerals are not pure substances, and are contaminated by whatever other elements are present in the given chemical system. As a result, it is possible for one element to be substituted for another. Chemical substitution will occur between ions of a similar size and charge; for example, K^+ will not substitute for Si^{4+} because of chemical and structural incompatibilities caused by a big difference in size and charge. A common example of chemical substitution is that of Si^{4+} by Al^{3+}, which are close in charge, size, and abundance in the crust. In the example of plagioclase, there are three cases of substitution. Feldspars are all framework silicates, which have a silicon-oxygen ratio of 2:1, and the space for other elements is given by the substitution of Si^{4+} by Al^{3+} to give a base unit of $[AlSi_3O_8]^-$; without the substitution, the formula would be charge-balanced as SiO_2, giving quartz. The significance of this structural property will be explained further by coordination polyhedra. The second substitution occurs between Na^+ and Ca^{2+}; however, the difference in charge has to accounted for by making a second substitution of Si^{4+} by Al^{3+}.

Coordination polyhedra are geometric representations of how a cation is surrounded by an anion. In mineralogy, coordination polyhedra are usually considered in terms of oxygen, due its abundance in the crust. The base unit of silicate minerals is the silica tetrahedron – one Si^{4+} surrounded by four O^{2-}. An alternate way of describing the coordination of the silicate is by a number: in the case of the silica tetrahedron, the silicon is said to have a coordination number of 4. Various cations have a specific range of possible coordination numbers; for silicon, it is almost always 4, except for very high-pressure minerals where the compound is compressed such that silicon is in six-fold (octahedral) coordination with oxygen. Bigger cations have a bigger coordination numbers because of the increase in relative size as compared to oxygen (the last orbital subshell of heavier atoms is different too). Changes in coordination numbers leads to physical and mineralogical differences; for example, at high pressure, such as in the mantle, many minerals, especially silicates such as olivine and garnet, will change to a perovskite structure, where silicon is in octahedral coordination. Other examples are the aluminosilicates kyanite, andalusite, and sillimanite (polymorphs, since they share the formula Al_2SiO_5), which differ by the coordination number of the Al^{3+}; these minerals transition from one another as a response to changes in pressure and temperature. In the case of silicate materials, the substitution of Si^{4+} by Al^{3+} allows for a variety of minerals because of the need to balance charges.

When minerals react, the products will sometimes assume the shape of the reagent; the product mineral is termed a pseudomorph of (or after) the reagent. Figure above is a pseudomorph of kaolinite after orthoclase. Here, the pseudomorph preserved the Carlsbad twinning common in orthoclase.

Changes in temperature and pressure and composition alter the mineralogy of a rock sample. Changes in composition can be caused by processes such as weathering or metasomatism (hydrothermal alteration). Changes in temperature and pressure occur when the host rock undergoes tectonic or magmatic movement into differing physical regimes. Changes in thermodynamic conditions make it favourable for mineral assemblages to react with each other to produce new minerals; as such, it is possible for two rocks to have an identical or a very similar bulk rock chemistry without having a similar mineralogy. This process of mineralogical alteration is related to the rock cycle. An example of a series of mineral reactions is illustrated as follows.

Orthoclase feldspar ($KAlSi_3O_8$) is a mineral commonly found in granite, a plutonic igneous rock. When exposed to weathering, it reacts to form kaolinite ($Al_2Si_2O_5(OH)_4$, a sedimentary mineral, and silicic acid):

$$2\,KAlSi_3O_8 + 5\,H_2O + 2\,H^+ \rightarrow Al_2Si_2O_5(OH)_4 + 4\,H_2SiO_3 + 2\,K^+$$

Under low-grade metamorphic conditions, kaolinite reacts with quartz to form pyrophyllite ($Al_2Si_4O_{10}(OH)_2$):

$$Al_2Si_2O_5(OH)_4 + SiO_2 \rightarrow Al_2Si_4O_{10}(OH)_2 + H_2O$$

As metamorphic grade increases, the pyrophyllite reacts to form kyanite and quartz:

$$Al_2Si_4O_{10}(OH)_2 \rightarrow Al_2SiO_5 + 3\ SiO_2 + H_2O$$

Alternatively, a mineral may change its crystal structure as a consequence of changes in temperature and pressure without reacting. For example, quartz will change into a variety of its SiO_2 polymorphs, such as tridymite and cristobalite at high temperatures, and coesite at high pressures.

Physical Properties

Classifying minerals ranges from simple to difficult. A mineral can be identified by several physical properties, some of them being sufficient for full identification without equivocation. In other cases, minerals can only be classified by more complex optical, chemical or X-ray diffraction analysis; these methods, however, can be costly and time-consuming. Physical properties applied for classification include crystal structure and habit, hardness, lustre, diaphaneity, colour, streak, cleavage and fracture, and specific gravity. Other less general tests include fluorescence, phosphorescence, magnetism, radioactivity, tenacity (response to mechanical induced changes of shape or form), piezoelectricity and reactivity to dilute acids.

Crystal Structure and Habit

Topaz has a characteristic orthorhombic elongated crystal shape.

Crystal structure results from the orderly geometric spatial arrangement of atoms in the internal structure of a mineral. This crystal structure is based on regular internal atomic or ionic arrangement that is often expressed in the geometric form that the crystal takes. Even when the mineral grains are too small to see or are irregularly shaped, the underlying crystal structure is always periodic and can be determined by X-ray diffraction. Minerals are typically described by their symmetry content. Crystals are restricted to 32 point groups, which differ by their symmetry. These groups are classified in turn into more broad categories, the most encompassing of these being the six crystal families.

These families can be described by the relative lengths of the three crystallographic axes, and the angles between them; these relationships correspond to the symmetry operations that define the narrower point groups. They are summarized below; a, b, and c represent the axes, and α, β, γ represent the angle opposite the respective crystallographic axis (e.g. α is the angle opposite the a-axis, viz. the angle between the b and c axes):

Crystal family	Lengths	Angles	Common examples
Isometric	$a=b=c$	$\alpha=\beta=\gamma=90°$	Garnet, halite, pyrite
Tetragonal	$a=b\neq c$	$\alpha=\beta=\gamma=90°$	Rutile, zircon, andalusite
Orthorhombic	$a\neq b\neq c$	$\alpha=\beta=\gamma=90°$	Olivine, aragonite, orthopyroxenes
Hexagonal	$a=b\neq c$	$\alpha=\beta=90°, \gamma=120°$	Quartz, calcite, tourmaline
Monoclinic	$a\neq b\neq c$	$\alpha=\gamma=90°, \beta\neq 90°$	Clinopyroxenes, orthoclase, gypsum
Triclinic	$a\neq b\neq c$	$\alpha\neq\beta\neq\gamma\neq 90°$	Anorthite, albite, kyanite

The hexagonal crystal family is also split into two crystal *systems* – the trigonal, which has a three-fold axis of symmetry, and the hexagonal, which has a six-fold axis of symmetry.

Chemistry and crystal structure together define a mineral. With a restriction to 32 point groups, minerals of different chemistry may have identical crystal structure. For example, halite (NaCl), galena (PbS), and periclase (MgO) all belong to the hexaoctahedral point group (isometric family), as they have a similar stoichiometry between their different constituent elements. In contrast, polymorphs are groupings of minerals that share a chemical formula but have a different structure. For example, pyrite and marcasite, both iron sulfides, have the formula FeS_2; however, the former is isometric while the latter is orthorhombic. This polymorphism extends to other sulfides with the generic AX_2 formula; these two groups are collectively known as the pyrite and marcasite groups.

Polymorphism can extend beyond pure symmetry content. The aluminosilicates are a group of three minerals – kyanite, andalusite, and sillimanite – which share the chemical formula Al_2SiO_5. Kyanite is triclinic, while andalusite and sillimanite are both orthorhombic and belong to the dipyramidal point group. These differences arise corresponding to how aluminium is coordinated within the crystal structure. In all minerals, one aluminium ion is always in six-fold coordination with oxygen. Silicon, as a general rule, is in four-fold coordination in all minerals; an exception is a case like stishovite (SiO_2, an ultra-high pressure quartz polymorph with rutile structure). In kyanite, the second aluminium is in six-fold coordination; its chemical formula can be expressed as $AlAlSiO_5$, to reflect its crystal structure. Andalusite has the second aluminium in five-fold coordination ($AlAlSiO_5$) and sillimanite has it in four-fold coordination ($AlAlSiO_5$).

Differences in crystal structure and chemistry greatly influence other physical properties of the mineral. The carbon allotropes diamond and graphite have vastly different properties; diamond is the hardest natural substance, has an adamantine lustre, and belongs to the isometric crystal family, whereas graphite is very soft, has a greasy lustre, and crystallises in the hexagonal family. This difference is accounted for by differences in bonding. In diamond, the carbons are in sp^3 hybrid orbitals, which means they form a framework where each carbon is covalently bonded to four neighbours in a tetrahedral fashion; on the other hand, graphite is composed of sheets of carbons in sp^2 hybrid orbitals, where each carbon is bonded covalently to only three others. These sheets are held together by much weaker van der Waals forces, and this discrepancy translates to large macroscopic differences.

Contact twins, as seen in spinel.

Twinning is the intergrowth of two or more crystals of a single mineral species. The geometry of the twinning is controlled by the mineral's symmetry. As a result, there are several types of twins, including contact twins, reticulated twins, geniculated twins, penetration twins, cyclic twins, and polysynthetic twins. Contact, or simple twins, consist of two crystals joined at a plane; this type of twinning is common in spinel. Reticulated twins, common in rutile, are interlocking crystals resembling netting. Geniculated twins have a bend in the middle that is caused by start of the twin. Penetration twins consist of two single crystals that have grown into each other; examples of this twinning include cross-shaped staurolite twins and Carlsbad twinning in orthoclase. Cyclic twins are caused by repeated twinning around a rotation axis. This type of twinning occurs around three, four, five, six, or eight-fold axes, and the corresponding patterns are called threelings, fourlings, fivelings, sixlings, and eightlings. Sixlings are common in aragonite. Polysynthetic twins are similar to cyclic twins through the presence of repetitive twinning; however, instead of occurring around a rotational axis, polysynthetic twinning occurs along parallel planes, usually on a microscopic scale.

Crystal habit refers to the overall shape of crystal. Several terms are used to describe this property. Common habits include acicular, which describes needlelike crystals as in natrolite, bladed, dendritic (tree-pattern, common in native copper), equant, which is typical of garnet, prismatic (elongated in one direction), and tabular, which differs from bladed habit in that the former is platy whereas the latter has a defined elongation. Related to crystal form, the quality of crystal faces is diagnostic of some minerals, especially with a petrographic microscope. Euhedral crystals have a defined external shape, while anhedral crystals do not; those intermediate forms are termed subhedral.

Hardness

The hardness of a mineral defines how much it can resist scratching. This physical property is controlled by the chemical composition and crystalline structure of a mineral. A mineral's hardness is not necessarily constant for all sides, which is a function of its structure; crystallographic weakness renders some directions softer than others. An example of this property exists in kyanite, which has a Mohs hardness of 5½ parallel to but 7 parallel to the most common scale of measurement is the ordinal Mohs hardness scale.

Diamond is the hardest natural material, and has a Mohs hardness of 10.

Defined by ten indicators, a mineral with a higher index scratches those below it. The scale ranges from talc, a phyllosilicate, to diamond, a carbon polymorph that is the hardest natural material. The scale is provided below:

Mohs hardness	Mineral	Chemical formula
1	Talc	$Mg_3Si_4O_{10}(OH)_2$
2	Gypsum	$CaSO_4{\cdot}2H_2O$
3	Calcite	$CaCO_3$
4	Fluorite	CaF_2
5	Apatite	$Ca_5(PO_4)_3(OH,Cl,F)$
6	Orthoclase	$KAlSi_3O_8$
7	Quartz	SiO_2
8	Topaz	$Al_2SiO_4(OH,F)_2$
9	Corundum	Al_2O_3
10	Diamond	C

Lustre and Diaphaneity

Lustre indicates how light reflects from the mineral's surface, with regards to its quality and intensity. There are numerous qualitative terms used to describe this property, which are split into metallic and non-metallic categories. Metallic and sub-metallic minerals have high reflectivity like metal; examples of minerals with this lustre are galena and pyrite. Non-metallic lustres include: adamantine, such as in diamond; vitreous, which is a glassy lustre very common in silicate minerals; pearly, such as in talc and apophyllite; resinous, such as members of the garnet group; silky which is common in fibrous minerals such as asbestiform chrysotile.

Pyrite has a metallic lustre.

The diaphaneity of a mineral describes the ability of light to pass through it. Transparent minerals do not diminish the intensity of light passing through them. An example of a transparent mineral is muscovite (potassium mica); some varieties are sufficiently clear to have been used for windows. Translucent minerals allow some light to pass, but less than those that are transparent. Jadeite and nephrite (mineral forms of jade are examples of minerals with this property). Minerals that do not allow light to pass are called opaque.

The diaphaneity of a mineral depends on the thickness of the sample. When a mineral is sufficiently thin (e.g., in a thin section for petrography), it may become transparent even if that property is not seen in a hand sample. In contrast, some minerals, such as hematite or pyrite, are opaque even in thin-section.

Colour and Streak

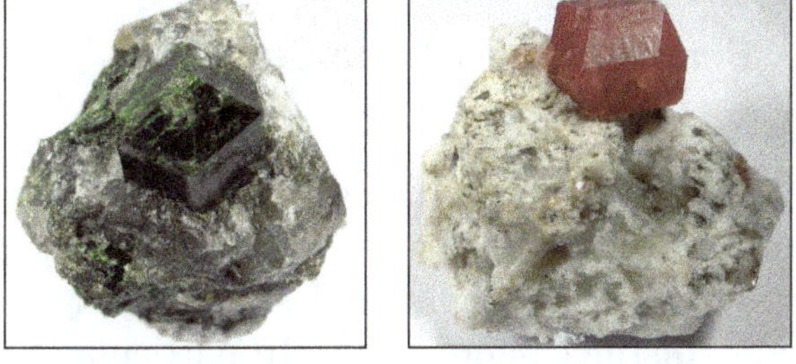

Colour is typically not a diagnostic property of minerals. Shown are green uvarovite (left) and red-pink grossular (right), both garnets. The diagnostic features would include dodecahedral crystals, resinous lustre, and hardness around 7.

Colour is the most obvious property of a mineral, but it is often non-diagnostic. It is caused by electromagnetic radiation interacting with electrons (except in the case of incandescence, which does not apply to minerals). Two broad classes of elements (idiochromatic and allochromatic) are defined with regards to their contribution to a mineral's colour: Idiochromatic elements are essential to a mineral's composition; their contribution to a mineral's colour is diagnostic. Examples of such minerals are malachite (green) and azurite (blue). In contrast, allochromatic elements in minerals are present in trace amounts as impurities. An example of such a mineral would be the ruby and sapphire varieties of the mineral corundum. The colours of pseudochromatic minerals are the result of interference of light waves. Examples include labradorite and bornite.

In addition to simple body colour, minerals can have various other distinctive optical properties, such as play of colours, asterism, chatoyancy, iridescence, tarnish, and pleochroism. Several of these properties involve variability in colour. Play of colour, such as in opal, results in the sample reflecting different colours as it is turned, while pleochroism describes the change in colour as light passes through a mineral in a different orientation. Iridescence is a variety of the play of colours where light scatters off a coating on the surface of crystal, cleavage planes, or off layers having minor gradations in chemistry. In contrast, the play of colours in opal is caused by light refracting from ordered microscopic silica spheres within its physical structure. Chatoyancy ("cat's eye") is the wavy banding of colour that is observed as the sample is rotated; asterism, a variety of chatoyancy, gives the appearance of a star on the mineral grain. The latter property is particularly common in gem-quality corundum.

The streak of a mineral refers to the colour of a mineral in powdered form, which may or may not be identical to its body colour. The most common way of testing this property is done with a streak plate, which is made out of porcelain and coloured either white or black. The streak of a mineral is independent of trace elements or any weathering surface. A common example of this property is illustrated with hematite, which is coloured black, silver, or red in hand sample, but has a cherry-red to reddish-brown streak. Streak is more often distinctive for metallic minerals, in contrast to non-metallic minerals whose body colour is created by allochromatic elements. Streak testing is constrained by the hardness of the mineral, as those harder than 7 powder the *streak plate* instead.

Cleavage, Parting, Fracture and Tenacity

By definition, minerals have a characteristic atomic arrangement. Weakness in this crystalline structure causes planes of weakness, and the breakage of a mineral along such planes is termed cleavage. The quality of cleavage can be described based on how cleanly and easily the mineral breaks; common descriptors, in order of decreasing quality, are "perfect", "good", "distinct", and "poor". In particularly transparent minerals, or in thin-section, cleavage can be seen as a series of parallel lines marking the planar surfaces when viewed from the side. Cleavage is not a universal property among minerals; for example, quartz, consisting of extensively interconnected silica tetrahedra, does not have a crystallographic weakness which would allow it to cleave. In contrast, micas, which have perfect basal cleavage, consist of sheets of silica tetrahedra which are very weakly held together.

Perfect basal cleavage as seen in biotite (black), and good cleavage seen in the matrix (pink orthoclase).

As cleavage is a function of crystallography, there are a variety of cleavage types. Cleavage occurs typically in either one, two, three, four, or six directions. Basal cleavage in one direction is a distinctive property of the micas. Two-directional cleavage is described as prismatic, and occurs in minerals such as the amphiboles and pyroxenes. Minerals such as galena or halite have cubic (or isometric) cleavage in three directions, at 90°; when three directions of cleavage are present, but not at 90°, such as in calcite or rhodochrosite, it is termed rhombohedral cleavage. Octahedral cleavage (four directions) is present in fluorite and diamond, and sphalerite has six-directional dodecahedral cleavage.

Minerals with many cleavages might not break equally well in all of the directions; for example, calcite has good cleavage in three directions, but gypsum has perfect cleavage in one direction, and poor cleavage in two other directions. Angles between cleavage planes vary between minerals. For example, as the amphiboles are double-chain silicates and the pyroxenes are single-chain silicates, the angle between their cleavage planes is different. The pyroxenes cleave in two directions at approximately 90°, whereas the amphiboles distinctively cleave in two directions separated by approximately 120° and 60°. The cleavage angles can be measured with a contact goniometer, which is similar to a protractor.

Parting, sometimes called "false cleavage", is similar in appearance to cleavage but is instead produced by structural defects in the mineral, as opposed to systematic weakness. Parting varies from crystal to crystal of a mineral, whereas all crystals of a given mineral will cleave if the atomic structure allows for that property. In general, parting is caused by some stress applied to a crystal. The sources of the stresses include deformation (e.g. an increase in pressure), exsolution, or twinning. Minerals that often display parting include the pyroxenes, hematite, magnetite, and corundum.

When a mineral is broken in a direction that does not correspond to a plane of cleavage, it is termed to have been fractured. There are several types of uneven fracture. The classic example is conchoidal fracture, like that of quartz; rounded surfaces are created, which are marked by smooth curved lines. This type of fracture occurs only in very homogeneous minerals. Other types of fracture are fibrous, splintery, and hackly. The latter describes a break along a rough, jagged surface; an example of this property is found in native copper.

Tenacity is related to both cleavage and fracture. Whereas fracture and cleavage describes the surfaces that are created when a mineral is broken, tenacity describes how resistant a mineral is to such breaking. Minerals can be described as brittle, ductile, malleable, sectile, flexible, or elastic.

Specific Gravity

Galena, PbS, is a mineral with a high specific gravity.

Specific gravity numerically describes the density of a mineral. The dimensions of density are mass divided by volume with units: kg/m^3 or g/cm^3. Specific gravity measures how much water a mineral sample displaces. Defined as the quotient of the mass of the sample and difference between the weight of the sample in air and its corresponding weight in water, specific gravity is a unitless ratio. Among most minerals, this property is not diagnostic. Rock forming minerals – typically silicates or occasionally carbonates – have a specific gravity of 2.5–3.5.

High specific gravity is a diagnostic property of a mineral. A variation in chemistry (and consequently, mineral class) correlates to a change in specific gravity. Among more common minerals, oxides and sulfides tend to have a higher specific gravity as they include elements with higher atomic mass. A generalization is that minerals with metallic or adamantine lustre tend to have higher specific gravities than those having a non-metallic to dull lustre. For example, hematite, Fe_2O_3, has a specific gravity of 5.26 while galena, PbS, has a specific gravity of 7.2–7.6, which is a result of their high iron and lead content, respectively. A very high specific gravity becomes very pronounced in native metals; kamacite, an iron-nickel alloy common in iron meteorites has a specific gravity of 7.9, and gold has an observed specific gravity between 15 and 19.3.

Other Properties

Carnotite (yellow) is a radioactive uranium-bearing mineral.

Other properties can be used to diagnose minerals. These are less general, and apply to specific minerals.

Dropping dilute acid (often 10% HCl) onto a mineral aids in distinguishing carbonates from other mineral classes. The acid reacts with the carbonate ($[CO_3]^{2-}$) group, which causes the affected area to effervesce, giving off carbon dioxide gas. This test can be further expanded to test the mineral in its original crystal form or powdered form. An example of this test is done when distinguishing calcite from dolomite, especially within the rocks (limestone and dolomite respectively). Calcite immediately effervesces in acid, whereas acid must be applied to powdered dolomite (often to a scratched surface in a rock), for it to effervesce. Zeolite minerals will not effervesce in acid; instead, they become frosted after 5–10 minutes, and if left in acid for a day, they dissolve or become a silica gel.

When tested, magnetism is a very conspicuous property of minerals. Among common minerals, magnetite exhibits this property strongly, and magnetism is also present, albeit not as strongly, in pyrrhotite and ilmenite. Some minerals exhibit electrical properties – for example, quartz is piezoelectric – but electrical properties are rarely used as diagnostic criteria for minerals because of incomplete data and natural variation.

Minerals can also be tested for taste or smell. Halite, NaCl, is table salt; its potassium-bearing counterpart, sylvite, has a pronounced bitter taste. Sulfides have a characteristic smell, especially as samples are fractured, reacting, or powdered.

Radioactivity is a rare property; minerals may be composed of radioactive elements. They could be a defining constituent, such as uranium in uraninite, autunite, and carnotite, or as trace impurities. In the latter case, the decay of a radioactive element damages the mineral crystal; the result, termed a *radioactive halo* or *pleochroic halo,* is observable with various techniques, such as thin-section petrography.

Classification

As the composition of the Earth's crust is dominated by silicon and oxygen, silicate elements are by far the most important class of minerals in terms of rock formation and diversity. However, non-silicate minerals are of great economic importance, especially as ores.

Non-silicate minerals are subdivided into several other classes by their dominant chemistry, which includes native elements, sulfides, halides, oxides and hydroxides, carbonates and nitrates, borates, sulfates, phosphates, and organic compounds. Most non-silicate mineral species are rare (constituting in total 8% of the Earth's crust), although some are relatively common, such as calcite, pyrite, magnetite, and hematite. There are two major structural styles observed in non-silicates: close-packing and silicate-like linked tetrahedra. close-packed structures is a way to densely pack atoms while minimizing interstitial space. Hexagonal close-packing involves stacking layers where every other layer is the same ("ababab"), whereas cubic close-packing involves stacking groups of three layers ("abcabcabc"). Analogues to linked silica tetrahedra include SO_4 (sulfate), PO_4 (phosphate), AsO_4 (arsenate), and VO_4 (vanadate). The non-silicates have great economic importance, as they concentrate elements more than the silicate minerals do.

The largest grouping of minerals by far are the silicates; most rocks are composed of greater than 95% silicate minerals, and over 90% of the Earth's crust is composed of these minerals. The two main constituents of silicates are silicon and oxygen, which are the two most abundant elements in the Earth's crust. Other common elements in silicate minerals correspond to other common elements in the Earth's crust, such as aluminium, magnesium, iron, calcium, sodium, and potassium. Some important rock-forming silicates include the feldspars, quartz, olivines, pyroxenes, amphiboles, garnets, and micas.

Silicates

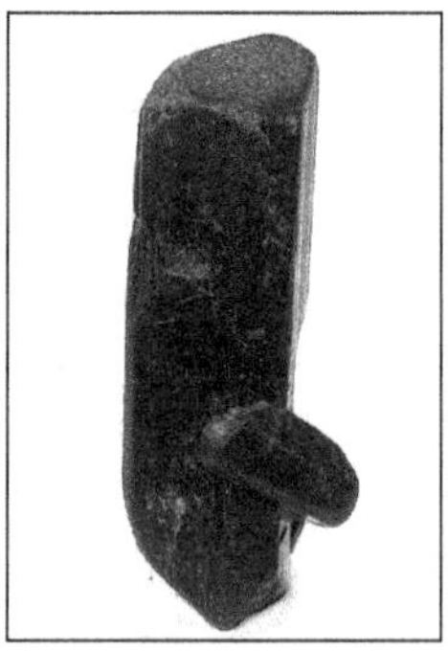

Aegirine, an iron-sodium clinopyroxene, is part of the inosilicate subclass.

The base unit of a silicate mineral is the $[SiO_4]^{4-}$ tetrahedron. In the vast majority of cases, silicon is in four-fold or tetrahedral coordination with oxygen. In very high-pressure situations, silicon will be in six-fold or octahedral coordination, such as in the perovskite structure or the quartz polymorph stishovite (SiO_2). In the latter case, the mineral no longer has a silicate structure, but that of rutile (TiO_2), and its associated group, which are simple oxides. These silica tetrahedra are then polymerized to some degree to create various structures, such as one-dimensional chains, two-dimensional sheets, and three-dimensional frameworks. The basic silicate mineral where no polymerization of the tetrahedra has occurred requires other elements to balance out the base 4- charge. In other silicate structures, different combinations of elements are required to balance out the resultant negative charge. It is common for the Si^{4+} to be substituted by Al^{3+} because of similarity in ionic radius and charge; in those cases, the $[AlO_4]^{5-}$ tetrahedra form the same structures as do the unsubstituted tetrahedra, but their charge-balancing requirements are different.

The degree of polymerization can be described by both the structure formed and how many tetrahedral corners (or coordinating oxygens) are shared (for aluminium and silicon in tetrahedral sites). Orthosilicates (or nesosilicates) have no linking of polyhedra, thus tetrahedra share no corners. Disilicates (or sorosilicates) have two tetrahedra sharing one oxygen atom. Inosilicates are chain silicates; single-chain silicates have two shared corners, whereas double-chain silicates have two or three shared corners. In phyllosilicates, a sheet structure is formed which requires three shared oxygens; in the case of double-chain silicates, some tetrahedra must share two corners instead of three as otherwise a sheet structure would result. Framework silicates, or tectosilicates, have tetrahedra that share all four corners. The ring silicates, or cyclosilicates, only need tetrahedra to share two corners to form the cyclical structure.

The silicate subclasses are described below in order of decreasing polymerization.

Tectosilicates

Natrolite is a mineral series in the zeolite group; this sample has a very prominent acicular crystal habit.

Tectosilicates, also known as framework silicates, have the highest degree of polymerization. With all corners of a tetrahedra shared, the silicon:oxygen ratio becomes 1:2. Examples are quartz, the feldspars, feldspathoids, and the zeolites. Framework silicates tend to be particularly chemically stable as a result of strong covalent bonds.

Forming 12% of the Earth's crust, quartz (SiO_2) is the most abundant mineral species. It is characterized by its high chemical and physical resistivity. Quartz has several polymorphs, including tridymite and cristobalite at high temperatures, high-pressure coesite, and ultra-high pressure stishovite. The latter mineral can only be formed on Earth by meteorite impacts, and its structure has been composed so much that it had changed from a silicate structure to that of rutile (TiO_2). The silica polymorph that is most stable at the Earth's surface is α-quartz. Its counterpart, β-quartz, is present only at high temperatures and pressures (changes to α-quartz below 573 °C at 1 bar). These two polymorphs differ by a "kinking" of bonds; this change in structure gives β-quartz greater symmetry than α-quartz, and they are thus also called high quartz (β) and low quartz (α).

Feldspars are the most abundant group in the Earth's crust, at about 50%. In the feldspars, Al^{3+} substitutes for Si^{4+}, which creates a charge imbalance that must be accounted for by the addition of cations. The base structure becomes either $[AlSi_3O_8]^-$ or $[Al_2Si_2O_8]^{2-}$. There are 22 mineral species of feldspars, subdivided into two major subgroups – alkali and plagioclase – and two less common groups – celsian and banalsite. The alkali feldspars are most commonly in a series between potassium-rich orthoclase and sodium-rich albite; in the case of plagioclase, the most common series ranges from albite to calcium-rich anorthite. Crystal twinning is common in feldspars, especially polysynthetic twins in plagioclase and Carlsbad twins in alkali feldspars. If the latter subgroup cools slowly from a melt, it forms exsolution lamellae because the two components – orthoclase and albite – are unstable in solid solution. Exsolution can be on a scale from microscopic to readily observable in hand-sample; perthitic texture forms when Na-rich feldspar exsolve in a K-rich host. The opposite texture (antiperthitic), where K-rich feldspar exsolves in a Na-rich host, is very rare.

Feldspathoids are structurally similar to feldspar, but differ in that they form in Si-deficient conditions, which allows for further substitution by Al^{3+}. As a result, feldspathoids cannot be associated with quartz. A common example of a feldspathoid is nepheline ($(Na, K)AlSiO_4$); compared to alkali feldspar, nepheline has an Al_2O_3:SiO_2 ratio of 1:2, as opposed to 1:6 in the feldspar. Zeolites often have distinctive crystal habits, occurring in needles, plates, or blocky masses. They form in the presence of water at low temperatures and pressures, and have channels and voids in their structure. Zeolites have several industrial applications, especially in waste water treatment.

Phyllosilicates

Muscovite, a mineral species in the mica group, within the phyllosilicate subclass.

Phyllosilicates consist of sheets of polymerized tetrahedra. They are bound at three oxygen sites, which gives a characteristic silicon:oxygen ratio of 2:5. Important examples include the mica, chlorite, and the kaolinite-serpentine groups. The sheets are weakly bound by van der Waals forces or hydrogen bonds, which causes a crystallographic weakness, in turn leading to a prominent basal cleavage among the phyllosilicates. In addition to the tetrahedra, phyllosilicates have a sheet of octahedra (elements in six-fold coordination by oxygen) that balance out the basic tetrahedra, which have a negative charge (e.g. $[Si_4O_{10}]^{4-}$). These tetrahedra (T) and octahedra (O) sheets are stacked in a variety of combinations to create phyllosilicate groups. Within an octahedral sheet, there are three octahedral sites in a unit structure; however, not all of the sites may be occupied. In that case, the mineral is termed dioctahedral, whereas in other case it is termed trioctahedral.

The kaolinite-serpentine group consists of T-O stacks (the 1:1 clay minerals); their hardness ranges from 2 to 4, as the sheets are held by hydrogen bonds. The 2:1 clay minerals (pyrophyllite-talc) consist of T-O-T stacks, but they are softer (hardness from 1 to 2), as they are instead held together by van der Waals forces. These two groups of minerals are subgrouped by octahedral occupation; specifically, kaolinite and pyrophyllite are dioctahedral whereas serpentine and talc trioctahedral.

Micas are also T-O-T-stacked phyllosilicates, but differ from the other T-O-T and T-O-stacked subclass members in that they incorporate aluminium into the tetrahedral sheets (clay minerals have Al^{3+} in octahedral sites). Common examples of micas are muscovite, and the biotite series. The chlorite group is related to mica group, but a brucite-like ($Mg(OH)_2$) layer between the T-O-T stacks.

Because of their chemical structure, phyllosilicates typically have flexible, elastic, transparent layers that are electrical insulators and can be split into very thin flakes. Micas can be used in electronics as insulators, in construction, as optical filler, or even cosmetics. Chrysotile, a species of serpentine, is the most common mineral species in industrial asbestos, as it is less dangerous in terms of health than the amphibole asbestos.

Inosilicates

Asbestiform tremolite, part of the amphibole group in the inosilicate subclass.

Inosilicates consist of tetrahedra repeatedly bonded in chains. These chains can be single, where a tetrahedron is bound to two others to form a continuous chain; alternatively, two chains can be merged to create double-chain silicates. Single-chain silicates have a silicon:oxygen ratio of 1:3 (e.g. $[Si_2O_6]^{4-}$), whereas the double-chain variety has a ratio of 4:11, e.g. $[Si_8O_{22}]^{12-}$. Inosilicates contain two important rock-forming mineral groups; single-chain silicates are most commonly pyroxenes, while double-chain silicates are often amphiboles. Higher-order chains exist (e.g. three-member, four-member, five-member chains, etc.) but they are rare.

The pyroxene group consists of 21 mineral species. Pyroxenes have a general structure formula of $XY(Si_2O_6)$, where X is an octahedral site, while Y can vary in coordination number from six to eight. Most varieties of pyroxene consist of permutations of Ca^{2+}, Fe^{2+} and Mg^{2+} to balance the negative charge on the backbone. Pyroxenes are common in the Earth's crust (about 10%) and are a key constituent of mafic igneous rocks.

Amphiboles have great variability in chemistry, described variously as a "mineralogical garbage can" or a "mineralogical shark swimming a sea of elements". The backbone of the amphiboles is the $[Si_8O_{22}]^{12-}$; it is balanced by cations in three possible positions, although the third position is not always used, and one element can occupy both remaining ones. Finally, the amphiboles are usually hydrated, that is, they have a hydroxyl group ($[OH]^-$), although it can be replaced by a fluoride, a chloride, or an oxide ion. Because of the variable chemistry, there are over 80 species of amphibole, although variations, as in the pyroxenes, most commonly involve mixtures of Ca^{2+}, Fe^{2+} and Mg^{2+}. Several amphibole mineral species can have an asbestiform crystal habit. These asbestos minerals form long, thin, flexible, and strong fibres, which are electrical insulators, chemically inert and heat-resistant; as such, they have several applications, especially in construction materials. However, asbestos are known carcinogens, and cause various other illnesses, such as asbestosis; amphibole asbestos (anthophyllite, tremolite, actinolite, grunerite, and riebeckite) are considered more dangerous than chrysotile serpentine asbestos.

Cyclosilicates

An example of elbaite, a species of tourmaline, with distinctive colour banding.

Cyclosilicates, or ring silicates, have a ratio of silicon to oxygen of 1:3. Six-member rings are most common, with a base structure of $[Si_6O_{18}]^{12-}$; examples include the tourmaline group and beryl. Other ring structures exist, with 3, 4, 8, 9, 12 having been described. Cyclosilicates tend to be strong, with elongated, striated crystals.

Tourmalines have a very complex chemistry that can be described by a general formula $XY_3Z_6(BO_3)_3T_6O_{18}V_3W$. The T_6O_{18} is the basic ring structure, where T is usually Si^{4+}, but substitutable by Al_{3+} or B^{3+}. Tourmalines can be subgrouped by the occupancy of the X site, and from there further subdivided by the chemistry of the W site. The Y and Z sites can accommodate a variety of cations, especially various transition metals; this variability in structural transition metal content gives the tourmaline group greater variability in colour. Other cyclosilicates include beryl, $Al_2Be_3Si_6O_{18}$, whose varieties include the gemstones emerald (green) and aquamarine (bluish). Cordierite is structurally similar to beryl, and is a common metamorphic mineral.

Sorosilicates

Epidote often has a distinctive pistachio-green colour.

Sorosilicates, also termed disilicates, have tetrahedron-tetrahedron bonding at one oxygen, which results in a 2:7 ratio of silicon to oxygen. The resultant common structural element is the $[Si_2O_7]^{6-}$ group. The most common disilicates by far are members of the epidote group. Epidotes are found in variety of geologic settings, ranging from mid-ocean ridge to granites to metapelites. Epidotes are built around the structure $[(SiO_4)(Si_2O_7)]^{10-}$ structure; for example, the mineral *species* epidote has calcium, aluminium, and ferric iron to charge balance: $Ca_2Al_2(Fe^{3+}, Al)(SiO_4)(Si_2O_7)O(OH)$. The presence of iron as Fe^{3+} and Fe^{2+} helps understand oxygen fugacity, which in turn is a significant factor in petrogenesis.

Other examples of sorosilicates include lawsonite, a metamorphic mineral forming in the blueschist facies (subduction zone setting with low temperature and high pressure), vesuvianite, which takes up a significant amount of calcium in its chemical structure.

Orthosilicates

Orthosilicates consist of isolated tetrahedra that are charge-balanced by other cations. Also termed nesosilicates, this type of silicate has a silicon:oxygen ratio of 1:4 (e.g.

SiO_4). Typical orthosilicates tend to form blocky equant crystals, and are fairly hard. Several rock-forming minerals are part of this subclass, such as the aluminosilicates, the olivine group, and the garnet group.

Black andradite, an end-member of the orthosilicate garnet group.

The aluminosilicates –bkyanite, andalusite, and sillimanite, all Al_2SiO_5 – are structurally composed of one $[SiO_4]^{4-}$ tetrahedron, and one Al^{3+} in octahedral coordination. The remaining Al^{3+} can be in six-fold coordination (kyanite), five-fold (andalusite) or four-fold (sillimanite); which mineral forms in a given environment is depend on pressure and temperature conditions. In the olivine structure, the main olivine series of (Mg, Fe)-$_2SiO_4$ consist of magnesium-rich forsterite and iron-rich fayalite. Both iron and magnesium are in octahedral by oxygen. Other mineral species having this structure exist, such as tephroite, Mn_2SiO_4. The garnet group has a general formula of $X_3Y_2(SiO_4)_3$, where X is a large eight-fold coordinated cation, and Y is a smaller six-fold coordinated cation. There are six ideal endmembers of garnet, split into two group. The pyralspite garnets have Al^{3+} in the Y position: pyrope ($Mg_3Al_2(SiO_4)_3$), almandine ($Fe_3Al_2(SiO_4)_3$), and spessartine ($Mn_3Al_2(SiO_4)_3$). The ugrandite garnets have Ca^{2+} in the X position: uvarovite ($Ca_3Cr_2(SiO_4)_3$), grossular ($Ca_3Al_2(SiO_4)_3$) and andradite ($Ca_3Fe_2(SiO_4)_3$). While there are two subgroups of garnet, solid solutions exist between all six end-members.

Other orthosilicates include zircon, staurolite, and topaz. Zircon ($ZrSiO_4$) is useful in geochronology as the Zr^{4+} can be substituted by U^{6+}; furthermore, because of its very resistant structure, it is difficult to reset it as a chronometer. Staurolite is a common metamorphic intermediate-grade index mineral. It has a particularly complicated crystal structure that was only fully described in 1986. Topaz ($Al_2SiO_4(F, OH)_2$, often found in granitic pegmatites associated with tourmaline, is a common gemstone mineral.

Non-silicates

Native Elements

Native elements are those that are not chemically bonded to other elements. This mineral group includes native metals, semi-metals, and non-metals, and various alloys and solid solutions. The metals are held together by metallic bonding, which confers distinctive physical properties such as their shiny metallic lustre, ductility and malleability, and electrical conductivity. Native elements are subdivided into groups by their structure or chemical attributes.

Native gold. Rare specimen of stout crystals growing off of a central stalk.

The gold group, with a cubic close-packed structure, includes metals such as gold, silver, and copper. The platinum group is similar in structure to the gold group. The iron-nickel group is characterized by several iron-nickel alloy species. Two examples are kamacite and taenite, which are found in iron meteorites; these species differ by the amount of Ni in the alloy; kamacite has less than 5–7% nickel and is a variety of native iron, whereas the nickel content of taenite ranges from 7–37%. Arsenic group minerals consist of semi-metals, which have only some metallic traits; for example, they lack the malleability of metals. Native carbon occurs in two allotropes, graphite and diamond; the latter forms at very high pressure in the mantle, which gives it a much stronger structure than graphite.

Sulfides

Red cinnabar (HgS), a mercury ore, on dolomite.

Sphalerite crystal partially encased in calcite from the Devonian Milwaukee Formation of Wisconsin.

The sulfide minerals are chemical compounds of one or more metals or semimetals with a sulfur; tellurium, arsenic, or selenium can substitute for the sulfur. Sulfides tend to be soft, brittle minerals with a high specific gravity. Many powdered sulfides, such as pyrite, have a sulfurous smell when powdered. Sulfides are susceptible to weathering, and many readily dissolve in water; these dissolved minerals can be later redeposited, which creates enriched secondary ore deposits. Sulfides are classified by the ratio of the metal or semimetal to the sulfur, such as M:S equal to 2:1, or 1:1. Many sulfide minerals are economically important as metal ores; examples include sphalerite (ZnS), an ore of zinc, galena (PbS), an ore of lead, cinnabar (HgS), an ore of mercury, and molybdenite (MoS_2, an ore of molybdenum. Pyrite (FeS_2), is the most commonly occurring sulfide, and can be found in most geological environments. It is not, however, an ore of iron, but can be instead oxidized to produce sulfuric acid. Related to the sulfides are the rare sulfosalts, in which a metallic element is bonded to sulfur and a semimetal such as antimony, arsenic, or bismuth. Like the sulfides, sulfosalts are typically soft, heavy, and brittle minerals.

Oxides

Oxide minerals are divided into three categories: simple oxides, hydroxides, and multiple oxides. Simple oxides are characterized by O^{2-} as the main anion and primarily ionic bonding. They can be further subdivided by the ratio of oxygen to the cations. The periclase group consists of minerals with a 1:1 ratio. Oxides with a 2:1 ratio include cuprite (Cu_2O) and water ice. Corundum group minerals have a 2:3 ratio, and includes minerals such as corundum (Al_2O_3), and hematite (Fe_2O_3). Rutile group minerals have a ratio of 1:2; the eponymous species, rutile (TiO_2) is the chief ore of titanium; other examples include cassiterite (SnO_2; ore of tin), and pyrolusite (MnO_2; ore of manganese). In hydroxides, the dominant anion is the hydroxyl ion, OH^-. Bauxites are the chief aluminium ore, and are a heterogeneous mixture of the hydroxide minerals diaspore, gibbsite, and bohmite; they form in areas with a very high rate of chemical weathering (mainly tropical conditions). Finally, multiple oxides are compounds of two metals with oxygen. A major group within this class are the spinels, with a general formula of $X^{2+}Y^{3+}{}_2O_4$. Examples of species include spinel ($MgAl_2O_4$), chromite ($FeCr_2O_4$), and magnetite (Fe_3O_4). The latter is readily distinguishable by its strong magnetism, which occurs as it has iron in two oxidation states ($Fe^{2+}Fe^{3+}{}_2O_4$), which makes it a multiple oxide instead of a single oxide.

Halides

Pink cubic halite (NaCl; halide class) crystals on a nahcolite matrix ($NaHCO_3$; a carbonate, and mineral form of sodium bicarbonate, used as baking soda).

The halide minerals are compounds in which a halogen (fluorine, chlorine, iodine, or bromine) is the main anion. These minerals tend to be soft, weak, brittle, and water-soluble. Common examples of halides include halite (NaCl, table salt), sylvite (KCl), fluorite (CaF_2). Halite and sylvite commonly form as evaporites, and can be dominant minerals in chemical sedimentary rocks. Cryolite, Na_3AlF_6, is a key mineral in the extraction of aluminium from bauxites; however, as the only significant occurrence at Ivittuut, Greenland, in a granitic pegmatite, was depleted, synthetic cryolite can be made from fluorite.

Carbonates

The carbonate minerals are those in which the main anionic group is carbonate, $[CO_3]^{2-}$. Carbonates tend to be brittle, many have rhombohedral cleavage, and all react with acid. Due to the last characteristic, field geologists often carry dilute hydrochloric acid to distinguish carbonates from non-carbonates. The reaction of acid with carbonates, most commonly found as the polymorph calcite and aragonite ($CaCO_3$), relates to the dissolution and precipitation of the mineral, which is a key in the formation of limestone caves, features within them such as stalactite and stalagmites, and karst landforms. Carbonates are most often formed as biogenic or chemical sediments in marine environments. The carbonate group is structurally a triangle, where a central C^{4+} cation is surrounded by three O^{2-} anions; different groups of minerals form from different arrangements of these triangles. The most common carbonate mineral is calcite, which is the primary constituent of sedimentary limestone and metamorphic marble. Calcite, $CaCO_3$, can have a high magnesium impurity. Under high-Mg conditions, its polymorph aragonite will form instead; the marine geochemistry in this regard can be described as an aragonite or calcite sea, depending on which mineral preferentially forms. Dolomite is a double carbonate, with the formula $CaMg(CO_3)_2$. Secondary dolomitization of limestone is common, in which calcite or aragonite are converted to dolomite; this reaction increases pore space (the unit cell volume of dolomite is 88% that of calcite), which can create a reservoir for oil and gas. These two mineral species are members of eponymous mineral groups: the calcite group includes carbonates with the general formula XCO_3, and the dolomite group constitutes minerals with the general formula $XY(CO_3)_2$.

Sulfates

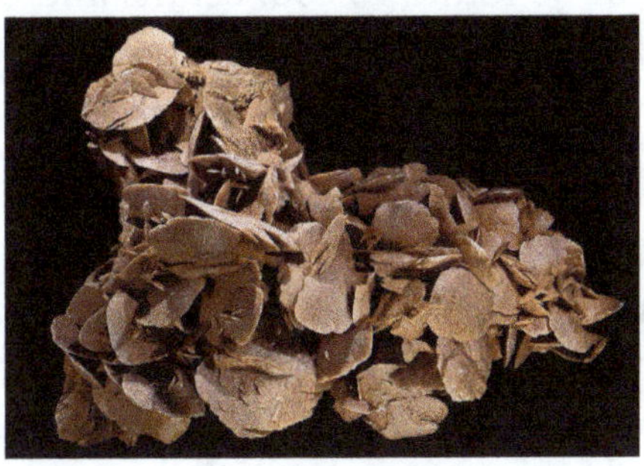

Gypsum desert rose.

The sulfate minerals all contain the sulfate anion, $[SO_4]^{2-}$. They tend to be transparent to translucent, soft, and many are fragile. Sulfate minerals commonly form as evaporites, where they precipitate out of evaporating saline waters. Sulfates can also be found in hydrothermal vein systems associated with sulfides, or as oxidation products of sulfides. Sulfates can be subdivided into anhydrous and hydrous minerals. The most common hydrous sulfate by far is gypsum, $CaSO_4 \cdot 2H_2O$. It forms as an evaporite, and is associated with other evaporites such as calcite and halite; if it incorporates sand grains as it crystallizes, gypsum can form desert roses. Gypsum has very low thermal conductivity and maintains a low temperature when heated as it loses that heat by dehydrating; as such, gypsum is used as an insulator in materials such as plaster and drywall. The anhydrous equivalent of gypsum is anhydrite; it can form directly from seawater in highly arid conditions. The barite group has the general formula XSO_4, where the X is a large 12-coordinated cation. Examples include barite ($BaSO_4$), celestine ($SrSO_4$), and anglesite ($PbSO_4$); anhydrite is not part of the barite group, as the smaller Ca^{2+} is only in eight-fold coordination.

Phosphates

The phosphate minerals are characterized by the tetrahedral $[PO_4]^{3-}$ unit, although the structure can be generalized, and phosphorus is replaced by antimony, arsenic, or vanadium. The most common phosphate is the apatite group; common species within this group are fluorapatite ($Ca_5(PO_4)_3F$), chlorapatite ($Ca_5(PO_4)_3Cl$) and hydroxylapatite ($Ca_5(PO_4)_3(OH)$). Minerals in this group are the main crystalline constituents of teeth and bones in vertebrates. The relatively abundant monazite group has a general structure of ATO_4, where T is phosphorus or arsenic, and A is often a rare-earth element (REE). Monazite is important in two ways: first, as a REE "sink", it can sufficiently concentrate these elements to become an ore; secondly, monazite group elements can incorporate relatively large amounts of uranium and thorium, which can be used in monazite geochronology to date the rock based on the decay of the U and Th to lead.

Organic Minerals

The Strunz classification includes a class for organic minerals. These rare compounds contain organic carbon, but can be formed by a geologic process. For example, whewellite, $CaC_2O_4 \cdot H_2O$ is an oxalate that can be deposited in hydrothermal ore veins. While hydrated calcium oxalate can be found in coal seams and other sedimentary deposits involving organic matter, the hydrothermal occurrence is not considered to be related to biological activity.

Astrobiology

It has been suggested that biominerals could be important indicators of extraterrestrial life and thus could play an important role in the search for past or present life on the planet Mars. Furthermore, organic components (biosignatures) that are often associated with biominerals are believed to play crucial roles in both pre-biotic and biotic reactions.

On January 24, 2014, NASA reported that current studies by the *Curiosity* and *Opportunity* rovers on Mars will now be searching for evidence of ancient life, including a biosphere based on autotrophic, chemotrophic and/or chemolithoautotrophic microorganisms, as well as ancient water, including fluvio-lacustrine environments (plains related to ancient rivers or lakes) that may have been habitable. The search for evidence of habitability, taphonomy (related to fossils), and organic carbon on the planet Mars is now a primary NASA objective.

Crystal Structure of Minerals

Usd axes are not compatible with crystallized materiad by its chemical composition, i.e. the kind of atoms and their ratios. This allows us to establish a chemical formula. However, neither the chemical compound nor the mineral is unequivocally defined. Even if both, the kind and the ratio of atoms as well as their relative positions, are known for a compound, the distinct phase is correctly characterized. Many minerals have the same chemical composition but exhibit different crystal structures.

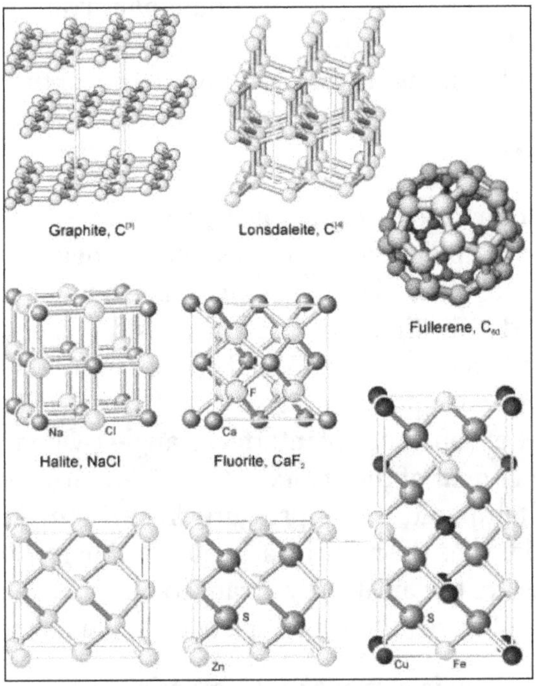

Some important crystal structures.

Even the variety of crystal structures formed only by one chemical element is conspicuous. As an example serves carbon: graphite and diamond are two well known minerals with quite different frequencies in occurrence, appearances, physical and chemical properties and commercial values; in addition, lonsdaleite is a rare mineral related to diamond but with another symmetry. During the last decades of the 20th century molecules formed by football-like arrangements of 60, 70, or more carbon atoms were found and named fullerenes (bucky balls). In the meantime, similarly constructed tube-like arranged carbon atoms were also recognized (bucky tubes). Obviously they are wide spread in soot. This shows that a chemical compound (or a mineral) is not correctly defined if the kind and ratio of chemical elements are known (as derived from a chemical analysis); it is also necessary to characterize the atomic arrangement, i.e. the crystal structure. In the past, an increasing number of crystal-structure investigations were performed by scattering experiments with X-ray, neutron or electron radiation.

The physical properties of a compound reflect at least some features of the atomic arrangement as well as the symmetry of the crystal structure. Often it is overseen that the types of the chemical bonds present within a crystal image physical properties such as luster, color and transparency. Etching figures, inclusions, anisotropy of hardness, cleavage, electrical and thermal conductivity, elasticity, rate of growth, piezoelectricity or pyroelectricity as well as other electric and magnetic properties also reflect the symmetry of a crystal and the character of its chemical bonds. The different symmetry and composition of crystals cause a distinct optical behavior of compounds as used for distinction of minerals with the help of a polarizing microscope. Optically positive and negative crystals are related to mainly axial and planar arranged chemical bonds. As a consequence, the characterization of the crystal symmetry and knowledge of the crystal structure is a useful tool for studying physical and chemical properties of minerals.

Symmetry Elements

Macroscopically, an undamaged crystal which grew unobstructed shows conspicuous regularities: it is formed by plane faces intersecting in edges. Often it is observed that two faces or some edges are parallel to each other. Faces and edges may be arranged in a distinct regularity. Already Nils Steensen, a Dane considered equal angles between analogue faces of quartz crystals in 1669. In 1688 these observations were extended to other crystal species by Domenico Guglielmini. It was Jean Bapstiste Louis Romé de l'Isle who proposed in 1783 the law of constancy of interfacial angles which states that the angle between two faces is independent of the crystal's size. Therefore the angles are maintained during crystal growth if temperature and pressure are unchanged. As a consequence, under defined physical conditions the interfacial angles are characteristic for any crystalline species. Faces and edges are not arbitrarily arranged but may be repeated by mathematically defined operations. The regular repetition is described by the symmetry elements.

The regular shape of a crystal is a result of the atomic arrangement. As a consequence, the crystal structure is reflected in the crystal=s habit. The habit of a crystal is its characteristic and common form defined by the regularly arranged faces. For many minerals, the crystal habit is so typical that it serves as a tool of identification. Already in the 18th century a crystal was thought to be formed from small sized regular bodies which are regularly arranged; they were called _molécule intégrante by Réné-Just Haüy. At least since experiments showed that X-rays are scattered by a crystal, the regular atomic construction of crystals was proved. The first X-ray scattering experiments were performed by Walter Friedrich and Paul Knipping under the supervision of Max von Laue in 1912.

For the morphological description and for description of the atomic arrangement of a crystal three non-planar (and general not orthogonal) vectors a, b, and c with lengths of a = |a|, b = |b|, and c = |c| are used. The vectors are chosen parallel to existing or at least possible edges of the crystal. The vectors are oriented in a right-handed coordinate system with the angles α (between b and c), β (between a and c), and γ (between a and b). The crystallographic axes of coordinates (labeled a, b, c or sometimes x, y, z) are chosen parallel to a, b, and c. The regularity of a crystal is characterized by symmetry elements. These are mathematical operations which describe the repetition of a crystal=s face, edge or corner (macroscopically) or an atom or molecule (on atomic scale) with respect to a point, a line or a plane. Moreover, symmetry elements can be applied to any motifs or objects. They are subdivided into "primary" and "secondary" symmetry elements. Primary symmetry elements are inversion, reflection or rotation. The successive execution (or the product) of two primary symmetry elements results in a new link between objects: One face or one atom or in general one motif is repeated after applying two different symmetry elements, i.e., a secondary symmetry element. Combinations of the symmetry elements are possible: A distinct number of primary and / or secondary symmetry elements are present which operate separately as if they solely existed. Symmetry elements which are used to describe the morphological habit of a crystal are also found within the atomic arrangement. However the macroscopically observable symmetry elements are either maintained or modified due to translation. In detail, the symmetry elements are defined as follows:

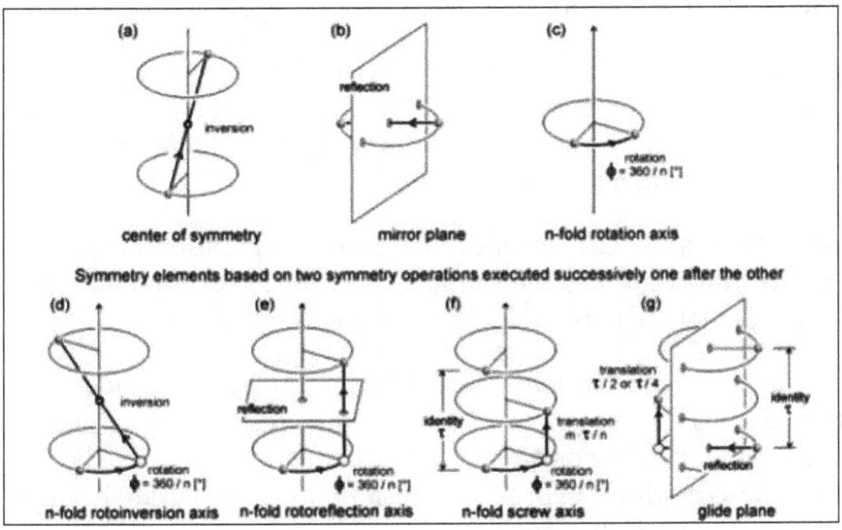

Repetition of two atoms according to symmetry elements. Symmetry elements based on one symmetry operation: (a) Center of symmetry (inversion center); (b) mirror plane; (c) rotation axis. Symmetry elements based on two symmetry operations executed successively one after the other (repetition according to a product of two symmetry elements): (d) rotoinversion axis (inversion axis); (e) rotoreflection axis; (f) screw axis; (g) glide plane.

- Identity (translation): A crystal is formed by a three-dimensional periodic lattice. Identical motifs (atoms, molecules) are repeated in equal amounts (translation period τ) in all directions of the three-dimensional space.

- Center of symmetry, also called an inversion center: This primary symmetry operation defines the repetition of the motif with respect to a point, i.e. the inversion center (symbol $\bar{1}$; manner of speaking >one bar=). An imaginary line can be passed from any point of a crystal, any atom in the crystal structure or any other object through the center of symmetry to an analogous point on the opposite side; the center of symmetry is located exactly in the middle between the symmetrically analogue pairs.

- Reflection plane, also called a mirror plane: This primary symmetry operation defines the repetition of a motif with respect to a plane (symbol m). A mirror plane is an imaginary plane passed through a crystal or any other object dividing it in half so that each half is the mirror image of the other.

- Rotation axis: This primary symmetry operation defines the repetition of a motif with respect to an n-fold rotation axis. After n times rotation of 360/n [°] identity of the motif is obtained, $n \in \mathbb{N}$ (symbol n). Rotation includes identity: It is obvious that an object is repeated after a rotation of 360/1 [°] into itself (one-fold axis, symbol 1). An axis of symmetry is an imaginary line through a crystal or any other object about which it may be rotated and repeated into itself after a complete revolution.

- Inversion axis, also called a axis of rotoinversion: This secondary symmetry operation defines the repetition of a motif with respect to the product of an n-fold rotation around an axis and an inversion at a point centered at this axis (symbol $\bar{n}$; manner of speaking >n bar=).

- Rotoreflection axis: This secondary symmetry operation defines the repetition of a motif with respect to the product of a rotation around an n-fold axis and a reflection at a plane perpendicular to this axis (symbol $\tilde{n}$).

- Screw axis, also called a axis of rototranslation: This secondary symmetry operation defines the repetition of a motif with respect to the product of a rotation around an n-fold axis and a translation parallel to this axis; the translation amounts $m \cdot \tau / n$ with τ being the translation period parallel to the n-fold axis and $1 \leq m < n$; $m, n \in \mathbb{N}$ (symbol n_m).

- Glide plane: This secondary symmetry operation defines the repetition of a motif with respect to the product of a reflection on a mirror plane and a translation (glide vector) τ/m parallel to this plane where m is 2 or 4. The translation vector is either parallel to a crystallographic axis, τ/m amounts a/2 (symbol a, a-glide plane), b/2 (symbol b, b-glide plane), c/2 (symbol c, c-glide plane), or it is parallel to the diagonal of a face of the unit cell, τ/m amounts (a+b)/2, (a+c)/2, (b+c)/2 (symbol n, nglide plane). In some F or I centered cells also glide planes with glide vectors of (a±b)/4, (a±c)/4, (b±c)/4, or (a±b±c)/4 occur (symbol d, d-glide plane).

There are some symmetry elements which describe the same repetition. The repetition of any motif by a one-fold rotoinversion $\overline{1}$ is identical to the description as a repetition of the motif by a center of symmetry. The two-fold rotoinversion $\overline{2}$ causes a repetition of a motif which is identical to the repetition of the motif by a mirror plane perpendicular to the $\overline{2}$ axis. The repetition by a rotoreflection may alternatively be described by a rotoinversion which, by convention, is used.

The application of a symmetry element results in two or more congruent objects. However, the congruence may either be direct or opposite. In the first case, conventional movement or rotation maintains the sign of corresponding angles within the object (transformation according to two right hands by a rotation, screw axes or translation) with the result that the two motifs are identical. In the second case, the two objects behave like left and right hands; they are enantiomorphous with respect to each other. The two objects may be transformed into each other by an inversion, a reflection, a rotoinversion, or a glide plane.

Macroscopically, only the center of symmetry, rotation axes, rotoinversion axes (rotoreflection axes), and mirror planes are observable on a crystal. The atomic arrangement exhibits the center of symmetry as the crystal macroscopically does. The rotation axes and rotoinversion axes are either maintained in the crystal structure or modified by a glide component (screw axes or glide planes). These are verified macroscopically without the glide component (i.e., as rotation axes and mirror planes). The reason is that the translation periods of crystals are roughly between 3 Å and 50 Å in minerals (the Ångström unit amounts to 1 Å = 10-10 m = 0.0001 µm = 0.1 nm = 100 pm; it is commonly used for the discussion of crystal structures notwithstanding that is no SI unit). With the best microscopes available we do not reach a magnification down to a level to image such a glide vector. There is a further essential difference between the symmetry of a crystal and that of the atomic arrangement: a crystal has only one symmetry center and neither collinear axes nor coplanar mirror planes occur. In contrast, the crystal structure exhibits series of symmetry elements arranged parallel to each other at equal distances due to translation.

Periodicity

Solid matter forming a three-dimensional periodic arrangement of atoms is called a 'crystal'. It is characterized by the periodic repetition of the motifs which are composed of atoms or molecules. Most minerals form crystals; there are only a few amorphic minerals. In amorphic compounds the atomic arrangement features only a short-range order, any longrange order is missing. One exception is naturally fused silica; the name of the mineral is lechatelierite. The silicon atoms are tetrahedrally coordinated by four oxygen atoms which are corner connected among each other, but neither the arrangement of the silicon atoms nor that of the oxygen atoms is periodic. Even calcedony or opal - often put to the amorphic minerals because viewable crystals are unknown - are at least in parts fine grained varieties of quartz, opal is sometimes mixed with cristobalite and / or tridymite.

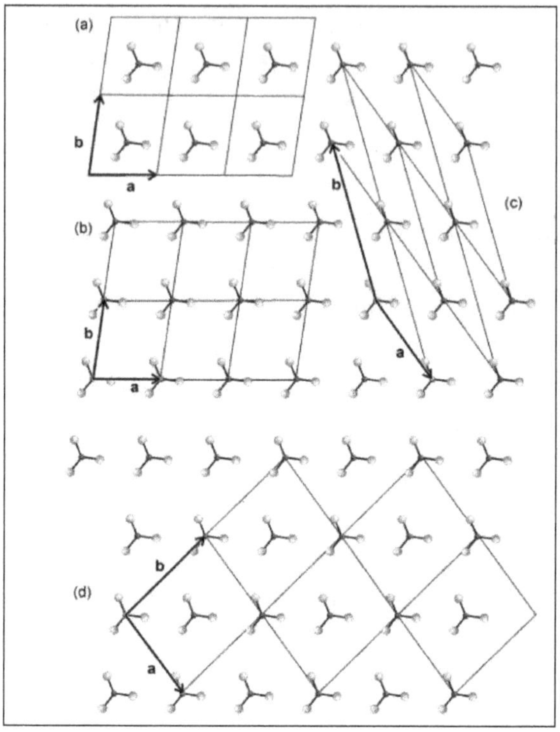

The repetition of a motif (a group of four atoms) in the two-dimensional space and four arbitrarily chosen translation lattices characterized by their unit cells. In (a) and (b) only the origin of the unit cell is shifted relative to each other. In (c) an uncommon (extremely flat) cell is represented. The lengths of the vectors **a** and b are unnecessarily long and define a so-called "not reduced cell". The unit cell chosen in (d) exhibits a doubled area which is unnecessarily large and involves two identical motifs. (It represents a centered cell.)

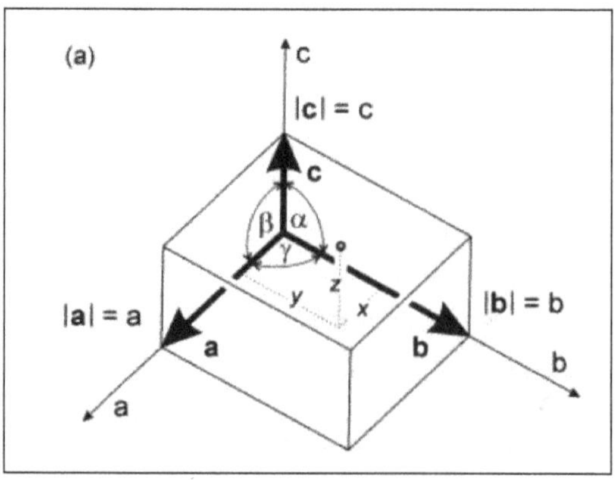

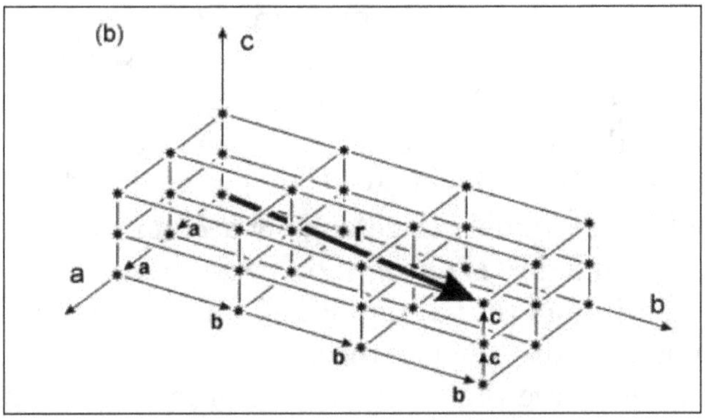

(a) A unit cell in the three-dimensional space characterized by the vectors **a**, **b**, and **c** parallel to the crystallographic axes a, b, and c. The lengths a, b, and c of the three vectors and the angles α, β, and γ between them are called "cell parameters". A point in the unit cell has the coordinates x, y, z. (b) A part of a three-dimensional lattice and a lattice vector **r** = **ua** + **vb** + **wc**, u, v, and $w \in \mathbb{N}$ (here [uvw] is [232]). The "lattice points" are marked by asterisks.

A periodic structure consists of motifs (atoms or molecules) which are repeated by translation. Infinity of the repetition within a crystal is assumed. The translation lattice (or short lattice) images the periodicity of the structure. Figure shows an example for the repetition of a motif within the two-dimensional space and four arbitrarily chosen lattices.

Each of these lattices defines a unit cell which may be used for the description of the same structure (i.e., the motif and the direction and amount of its translation). It is to be considered that the lattice does not define the arrangement of atoms within the unit cell. In the two-dimensional space, the translation lattice is characterized by a parallelogram defined by two non-collinear translation vectors a and b. In the three-dimensional space the translation lattice is characterized by a parallelepiped defined by three non-coplanar vectors a, b, and c where a = |a|, b = |b| and c = |c| and the angles α, β, and γ between them are called 'cell parameters'.

The parallelogram as well as the parallelepiped is called "unit cell" or simple "cell". The corners of the unit cell are called "lattice points". The lattice points do not necessarily represent atoms or other physical objects; i.e., the origin of the lattice is not necessarily in an atom or in the center of an atomic group; it can be chosen arbitrarily. In two-dimensional space the lattice points are described by the translation vectors.

$$\mathbf{r} = \mathrm{u}\,\mathbf{a} + \mathrm{v}\,\mathbf{b}, \quad \mathrm{u\,and\,v} \in \mathbb{N};$$

in three-dimensional space they are described by:

$$\mathbf{r} = \mathrm{u}\,\mathbf{a} + \mathrm{v}\,\mathbf{b} + wc, \quad \mathrm{u}, v \,\mathrm{and}\; w \in \mathbb{N};$$

Note that [uvw] is called the "zone" (or more precisely zone axis) of a crystal which denotes the direction of an existing or possible edge of the crystal. However, there are international conventions for the choice of the unit cell with respect to the angles α, β, and γ (and the length of the edges of the unit cell). An example is shown in figure. The common (reduced) cell for the lattice shown is that in figure. The cell selected in figure is unusual because of the extremely large angle γ accompanied by unnecessary large vectors a and b. The cell in figure is uncommon because its area is unnecessarily large since there are two identical motifs within the cell. In addition, there are international conventions for the choice of the origin of the lattice with respect to symmetry elements if present (as a reference serve the "International Tables for Crystallography"). The coordinates x, y, and z of a point within the unit cell are referred to this non-unitary coordinate system and $0 \leq x, y, z < 1$. Thus the coordinates of an atom or of any point within the unit cell are dimension-less numbers.

In this context it is worthy to note that the principle of translation allows only one-, two-, three-, four-, and six-fold axes. In figures it is shown that the two-dimensional space can be filled with triangles, the parallelogram (including rectangles and squares) and hexagons. In contrast, regular penta-, hepta-, octa-, nona-, deca-, ... -gons cannot fill planar space solely. As a consequence, five-, seven-, eight-, nine-, ten-, ... -fold axes are not compatible with crystallized materials.

Ceramics

A ceramic is a solid material comprising an inorganic compound of metal, non-metal or ionic and covalent bonds. Common examples are earthenware, porcelain, and brick.

The crystallinity of ceramic materials ranges from highly oriented to semi-crystalline, vitrified, and often completely amorphous (e.g., glasses). Most often, fired ceramics are either vitrified or semi-vitrified as is the case with earthenware, stoneware, and porcelain. Varying crystallinity and electron composition in the ionic and covalent bonds cause most ceramic materials to be good thermal and electrical insulators (extensively researched in ceramic engineering). With such a large range of possible options for the composition/structure of a ceramic (e.g. nearly all of the elements, nearly all types of bonding, and all levels of crystallinity), the breadth of the subject is vast, and identifiable attributes (e.g. hardness, toughness, electrical conductivity, etc.) are difficult to specify for the group as a whole. General properties such as high melting temperature, high hardness, poor conductivity, high moduli of elasticity, chemical resistance and low ductility are the norm, with known exceptions to each of these rules (e.g. piezoelectric ceramics, glass transition temperature, superconductive ceramics, etc). Many composites, such as fiberglass and carbon fiber, while containing ceramic materials, are not considered to be part of the ceramic family.

The earliest ceramics made by humans were pottery objects (i.e. *pots* or *vessels*) or figurines made from clay, either by itself or mixed with other materials like silica, hardened and sintered in fire. Later ceramics were glazed and fired to create smooth, colored surfaces, decreasing porosity through the use of glassy, amorphous ceramic coatings on top of the crystalline ceramic substrates. Ceramics now include domestic, industrial and building products, as well as a wide range of ceramic art. In the 20th century, new ceramic materials were developed for use in advanced ceramic engineering, such as in semiconductors.

Short timeline of ceramic in different styles.

Materials

A ceramic material is an inorganic, non-metallic, often crystalline oxide, nitride or carbide material. Some elements, such as carbon or silicon, may be considered ceramics.

Ceramic materials are brittle, hard, strong in compression, and weak in shearing and tension. They withstand chemical erosion that occurs in other materials subjected to acidic or caustic environments. Ceramics generally can withstand very high temperatures, ranging from 1,000 °C to 1,600 °C (1,800 °F to 3,000 °F). Glass is often not considered a ceramic because of its amorphous (noncrystalline) character. However, glassmaking involves several steps of the ceramic process, and its mechanical properties are similar to ceramic materials.

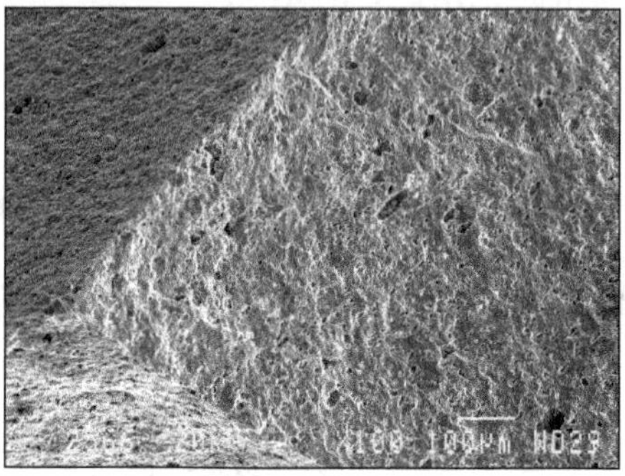

A low magnification SEM micrograph of an advanced ceramic material. The properties of ceramics make fracturing an import ant inspection method.

Traditional ceramic raw materials include clay minerals such as kaolinite, whereas more recent materials include aluminium oxide, more commonly known as alumina. The modern ceramic materials, which are classified as advanced ceramics, include silicon carbide and tungsten carbide. Both are valued for their abrasion resistance and hence find use in applications such as the wear plates of crushing equipment in mining operations. Advanced ceramics are also used in the medicine, electrical, electronics industries and body armor.

Crystalline Ceramics

Crystalline ceramic materials are not amenable to a great range of processing. Methods for dealing with them tend to fall into one of two categories – either make the ceramic in the desired shape, by reaction *in situ*, or by "forming" powders into the desired shape, and then sintering to form a solid body. Ceramic forming techniques include shaping by hand (sometimes including a rotation process called "throwing"), slip casting, tape casting (used for making very thin ceramic capacitors), injection molding, dry pressing, and other variations.

Noncrystalline Ceramics

Noncrystalline ceramics, being glass, tend to be formed from melts. The glass is shaped when either fully molten, by casting, or when in a state of toffee-like viscosity, by methods such as blowing into a mold. If later heat treatments cause this glass to become partly crystalline, the resulting material is known as a glass-ceramic, widely used as cook-tops and also as a glass composite material for nuclear waste disposal.

Properties

The physical properties of any ceramic substance are a direct result of its crystalline structure and chemical composition. Solid-state chemistry reveals the fundamental connection between microstructure and properties such as localized density variations, grain size distribution, type of porosity and second-phase content, which can all be correlated with ceramic properties such as mechanical strength σ by the Hall-Petch equation, hardness, toughness, dielectric constant, and the optical properties exhibited by transparent materials.

Ceramography is the art and science of preparation, examination and evaluation of ceramic microstructures. Evaluation and characterization of ceramic microstructures is often implemented on similar spatial scales to that used commonly in the emerging field of nanotechnology: from tens of angstroms (A) to tens of micrometers (μm). This is typically somewhere between the minimum wavelength of visible light and the resolution limit of the naked eye.

The microstructure includes most grains, secondary phases, grain boundaries, pores, micro-cracks, structural defects and hardness microindentions. Most bulk mechanical, optical, thermal, electrical and magnetic properties are significantly affected by the observed microstructure. The fabrication method and process conditions are generally indicated by the microstructure. The root cause of many ceramic failures is evident in the cleaved and polished microstructure. Physical properties which constitute the field of materials science and engineering include the following:

Mechanical Properties

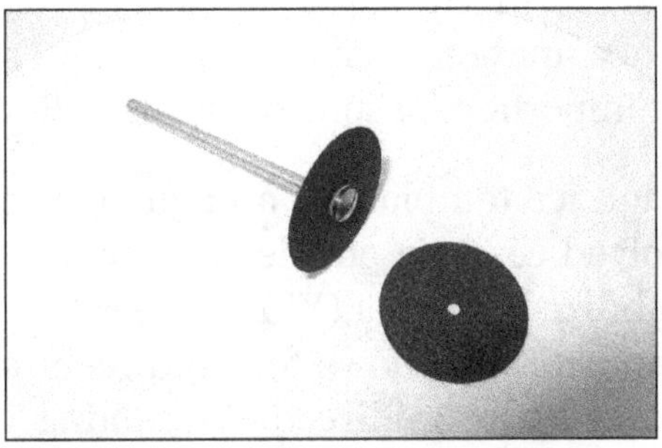

Cutting disks made of silicon carbide.

Mechanical properties are important in structural and building materials as well as textile fabrics. In modern materials science, fracture mechanics is an important tool in improving the mechanical performance of materials and components. It applies the physics of stress and strain, in particular the theories of elasticity and plasticity, to the microscopic crystallographic defects found in real materials in order to predict the macroscopic mechanical failure of bodies. Fractography is widely used with fracture mechanics to understand the causes of failures and also verify the theoretical failure predictions with real life failures.

Ceramic materials are usually ionic or covalent bonded materials, and can be crystalline or amorphous. A material held together by either type of bond will tend to fracture before any plastic deformation takes place, which results in poor toughness in these materials. Additionally, because these materials tend to be porous, the pores and other microscopic imperfections act as stress concentrators, decreasing the toughness further, and reducing the tensile strength. These combine to give catastrophic failures, as opposed to the more ductile failure modes of metals.

These materials do show plastic deformation. However, because of the rigid structure of the crystalline materials, there are very few available slip systems for dislocations to move, and so they deform very slowly. With the non-crystalline (glassy) materials, viscous flow is the dominant source of plastic deformation, and is also very slow. It is therefore neglected in many applications of ceramic materials.

To overcome the brittle behaviour, ceramic material development has introduced the class of ceramic matrix composite materials, in which ceramic fibers are embedded and with specific coatings are forming fiber bridges across any crack. This mechanism substantially increases the fracture toughness of such ceramics. Ceramic disc brakes are an example of using a ceramic matrix composite material manufactured with a specific process.

Ice-templating for Enhanced Mechanical Properties

If a ceramic will be subjected to substantial mechanical loading it can undergo a process called ice-templating, which allows some control of the microstructure of the ceramic product and therefore some control of the mechanical properties. Ceramic engineers use this technique to tune the mechanical properties to their desired application. Specifically, strength is increased when this technique is employed. Ice templating allows the creation of macroscopic pores in a unidirectional arrangement. The applications of this oxide strengthening technique are important for solid oxide fuel cells and water filtration devices.

To process a sample through ice templating, an aqueous colloidal suspension is prepared containing the dissolved ceramic powder evenly dispersed throughout the colloid, for example Yttria-stabilized zirconia (YSZ). The solution is then cooled from the bottom to the top on a platform that allows for unidirectional cooling. This forces ice crystals to grow in compliance to the unidirectional cooling, and these ice crystals force the dissolved YSZ particles to the solidification front of the solid-liquid interphase boundary, resulting in pure ice crystals lined up unidirectionally alongside concentrated pockets of colloidal particles. The sample is then simultaneously heated and the pressure is reduced enough to force the ice crystals to sublimate and the YSZ pockets begin to anneal together to form macroscopically aligned ceramic microstuctures. The sample is then further sintered to complete the evaporation of the residual water and the final consolidation of the ceramic microstructure.

During ice-templating a few variables can be controlled to influence the pore size and morphology of the microstructure. These important variables are the initial solids loading of the colloid, the cooling rate, the sintering temperature and duration, and the use of certain additives which can influence the micro-structural morphology during the process. A good understanding of these parameters is essential to understanding the relationships between processing, microstructure, and mechanical properties of anisotropically porous materials.

Electrical Properties

Semiconductors

Some ceramics are semiconductors. Most of these are transition metal oxides that are II-VI semiconductors, such as zinc oxide.

While there are prospects of mass-producing blue LEDs from zinc oxide, ceramicists are most interested in the electrical properties that show grain boundary effects.

One of the most widely used of these is the varistor. These are devices that exhibit the property that resistance drops sharply at a certain threshold voltage. Once the voltage across the device reaches the threshold, there is a breakdown of the electrical structure in the vicinity of the grain boundaries, which results in its electrical resistance dropping from several megohms down to a few hundred ohms. The major advantage of these is that they can dissipate a lot of energy, and they self-reset – after the voltage across the device drops below the threshold, its resistance returns to being high.

This makes them ideal for surge-protection applications; as there is control over the threshold voltage and energy tolerance, they find use in all sorts of applications. The best demonstration of their ability can be found in electrical substations, where they are employed to protect the infrastructure from lightning strikes. They have rapid response, are low maintenance, and do not appreciably degrade from use, making them virtually ideal devices for this application.

Semiconducting ceramics are also employed as gas sensors. When various gases are passed over a polycrystalline ceramic, its electrical resistance changes. With tuning to the possible gas mixtures, very inexpensive devices can be produced.

Superconductivity

The Meissner effect demonstrated by levitating a magnet above a cuprate superconductor, which is cooled by liquid nitrogen.

Under some conditions, such as extremely low temperature, some ceramics exhibit high-temperature superconductivity. The reason for this is not understood, but there are two major families of superconducting ceramics.

Ferroelectricity and Supersets

Piezoelectricity, a link between electrical and mechanical response, is exhibited by a large number of ceramic materials, including the quartz used to measure time in watches and other electronics. Such devices use both properties of piezoelectrics, using electricity to produce a mechanical motion (powering the device) and then using this mechanical motion to produce electricity (generating a signal). The unit of time measured is the natural interval required for electricity to be converted into mechanical energy and back again.

The piezoelectric effect is generally stronger in materials that also exhibit pyroelectricity, and all pyroelectric materials are also piezoelectric. These materials can be used to inter-convert between thermal, mechanical, or electrical energy; for instance, after synthesis in a furnace, a pyroelectric crystal allowed to cool under no applied stress generally builds up a static charge of thousands of volts. Such materials are used in motion sensors, where the tiny rise in temperature from a warm body entering the room is enough to produce a measurable voltage in the crystal.

In turn, pyroelectricity is seen most strongly in materials which also display the ferroelectric effect, in which a stable electric dipole can be oriented or reversed by applying an electrostatic field. Pyroelectricity is also a necessary consequence of ferroelectricity. This can be used to store information in ferroelectric capacitors, elements of ferroelectric RAM.

The most common such materials are lead zirconate titanate and barium titanate. Aside from the uses mentioned above, their strong piezoelectric response is exploited in the design of high-frequency loudspeakers, transducers for sonar, and actuators for atomic force and scanning tunneling microscopes.

Positive Thermal Coefficient

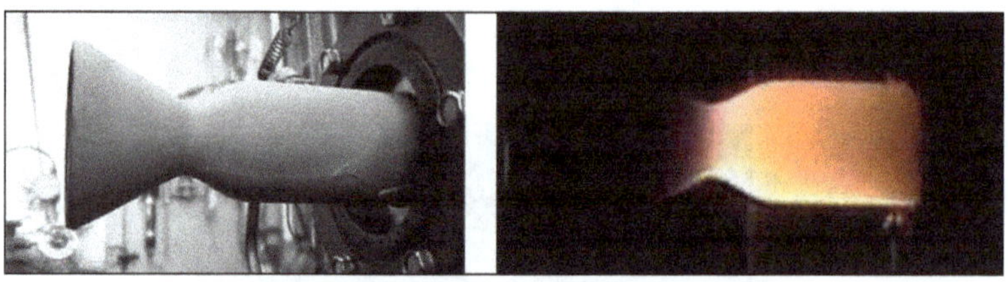

Silicon nitride rocket thruster. Left: Mounted in test stand.
Right: Being tested with H_2/O_2 propellants.

Increases in temperature can cause grain boundaries to suddenly become insulating in some semiconducting ceramic materials, mostly mixtures of heavy metal titanates. The critical transition temperature can be adjusted over a wide range by variations in chemistry. In such materials, current will pass through the material until joule heating brings it to the transition temperature, at which point the circuit will be broken and current flow will cease. Such ceramics are used as self-controlled heating elements in, for example, the rear-window defrost circuits of automobiles.

At the transition temperature, the material's dielectric response becomes theoretically infinite. While a lack of temperature control would rule out any practical use of the material near its critical temperature, the dielectric effect remains exceptionally strong even at much higher temperatures. Titanates with critical temperatures far below room temperature have become synonymous with "ceramic" in the context of ceramic capacitors for just this reason.

Optical Properties

Cermax xenon arc lamp with synthetic sapphire output window.

Optically transparent materials focus on the response of a material to incoming lightwaves of a range of wavelengths. Frequency selective optical filters can be utilized to alter or enhance the brightness and contrast of a digital image. Guided lightwave transmission via frequency selective waveguides involves the emerging field of fiber optics and the ability of certain glassy compositions as a transmission medium for a range of frequencies simultaneously (multi-mode optical fiber) with little or no interference between competing wavelengths or frequencies. This resonant mode of energy and data transmission via electromagnetic (light) wave propagation, though low powered, is virtually lossless. Optical waveguides are used as components in Integrated optical circuits (e.g. light-emitting diodes, LEDs) or as the transmission medium in local and long haul optical communication systems. Also of value to the emerging materials scientist is the sensitivity of materials to radiation in the thermal infrared (IR) portion of the electromagnetic spectrum. This heat-seeking ability is responsible for such diverse optical phenomena as Night-vision and IR luminescence.

Thus, there is an increasing need in the military sector for high-strength, robust materials which have the capability to transmit light (electromagnetic waves) in the visible (0.4 – 0.7 micrometers) and mid-infrared (1 – 5 micrometers) regions of the spectrum. These materials are needed for applications requiring transparent armor, including next-generation high-speed missiles and pods, as well as protection against improvised explosive devices (IED).

In the 1960s, scientists at General Electric (GE) discovered that under the right manufacturing conditions, some ceramics, especially aluminium oxide (alumina), could be made translucent. These translucent materials were transparent enough to be used for containing the electrical plasma generated in high-pressure sodium street lamps. During the past two decades, additional types of transparent ceramics have been developed for applications such as nose cones for heat-seeking missiles, windows for fighter aircraft, and scintillation counters for computed tomography scanners.

In the early 1970s, Thomas Soules pioneered computer modeling of light transmission through translucent ceramic alumina. His model showed that microscopic pores in ceramic, mainly trapped at the junctions of microcrystalline grains, caused light to scatter and prevented true transparency. The volume fraction of these microscopic pores had to be less than 1% for high-quality optical transmission.

This is basically a particle size effect. Opacity results from the incoherent scattering of light at surfaces and interfaces. In addition to pores, most of the interfaces in a typical metal or ceramic object are in the form of grain boundaries which separate tiny regions of crystalline order. When the size of the scattering center (or grain boundary) is reduced below the size of the wavelength of the light being scattered, the scattering no longer occurs to any significant extent.

In the formation of polycrystalline materials (metals and ceramics) the size of the crystalline grains is determined largely by the size of the crystalline particles present in the raw material during formation (or pressing) of the object. Moreover, the size of the grain boundaries scales directly with particle size. Thus a reduction of the original particle size below the wavelength of visible light (~ 0.5 micrometers for shortwave violet) eliminates any light scattering, resulting in a transparent material.

Recently, Japanese scientists have developed techniques to produce ceramic parts that rival the transparency of traditional crystals (grown from a single seed) and exceed the fracture toughness of a single crystal. In particular, scientists at the Japanese firm Konoshima Ltd., a producer of ceramic construction materials and industrial chemicals, have been looking for markets for their transparent ceramics.

Livermore researchers realized that these ceramics might greatly benefit high-powered lasers used in the National Ignition Facility (NIF) Programs Directorate. In particular, a Livermore research team began to acquire advanced transparent ceramics from Konoshima to determine if they could meet the optical requirements needed for Livermore's Solid-State Heat Capacity Laser (SSHCL). Livermore researchers have also been testing applications of these materials for applications such as advanced drivers for laser-driven fusion power plants.

Examples:

Porcelain high-voltage insulator.

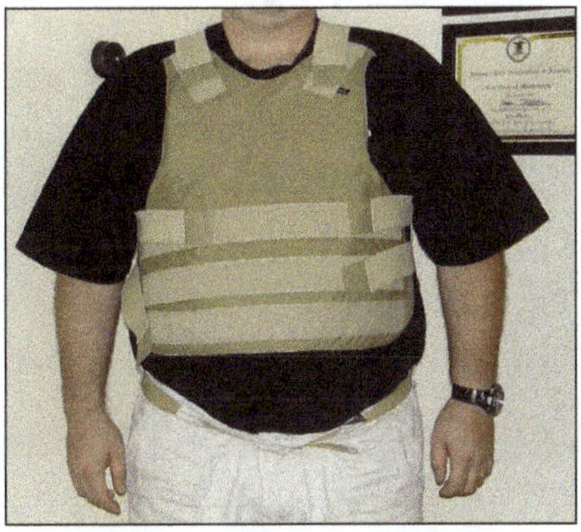

Silicon carbide is used for inner plates of ballistic vests.

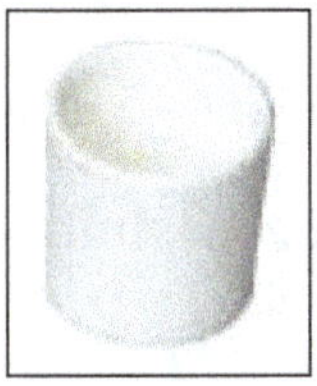

Ceramic BN crucible.

A composite material of ceramic and metal is known as cermet.

Other ceramic materials, generally requiring greater purity in their make-up than those above, include forms of several chemical compounds, including:

- Barium titanate (often mixed with strontium titanate) displays ferroelectricity, meaning that its mechanical, electrical, and thermal responses are coupled to one another and also history-dependent. It is widely used in electromechanical transducers, ceramic capacitors, and data storage elements. Grain boundary conditions can create PTC effects in heating elements.
- Bismuth strontium calcium copper oxide, a high-temperature superconductor.
- Boron oxide is used in body armour.
- Boron nitride is structurally isoelectronic to carbon and takes on similar physical forms: a graphite-like one used as a lubricant, and a diamond-like one used as an abrasive.
- Earthenware used for domestic ware such as plates and mugs.
- Ferrite is used in the magnetic cores of electrical transformers and magnetic core memory.
- Lead zirconate titanate (PZT) was developed at the United States National

Bureau of Standards in 1954. PZT is used as an ultrasonic transducer, as its piezoelectric properties greatly exceed those of Rochelle salt.

- Magnesium diboride (MgB_2) is an unconventional superconductor.
- Porcelain is used for a wide range of household and industrial products.
- Sialon (Silicon Aluminium Oxynitride) has high strength; resistance to thermal shock, chemical and wear resistance, and low density. These ceramics are used in non-ferrous molten metal handling, weld pins and the chemical industry.
- Silicon carbide (SiC) is used as a susceptor in microwave furnaces, a commonly used abrasive, and as a refractory material.
- Silicon nitride (Si_3N_4) is used as an abrasive powder.
- Steatite (magnesium silicates) is used as an electrical insulator.
- Titanium carbide used in space shuttle re-entry shields and scratchproof watches.
- Uranium oxide (UO_2), used as fuel in nuclear reactors.
- Yttrium barium copper oxide ($YBa_2Cu_3O_{7-x}$), another high temperature superconductor.
- Zinc oxide (ZnO), which is a semiconductor, and used in the construction of varistors.
- Zirconium dioxide (zirconia), which in pure form undergoes many phase changes between room temperature and practical sintering temperatures, can be chemically "stabilized" in several different forms. Its high oxygen ion conductivity recommends it for use in fuel cells and automotive oxygen sensors. In another variant, metastable structures can impart transformation toughening for mechanical applications; most ceramic knife blades are made of this material.
- Partially stabilised zirconia (PSZ) is much less brittle than other ceramics and is used for metal forming tools, valves and liners, abrasive slurries, kitchen knives and bearings subject to severe abrasion.

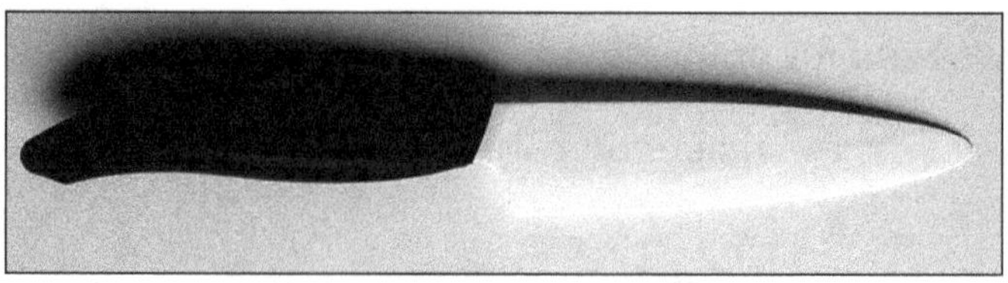

Kitchen knife with a ceramic blade.

Products

By Usage

For convenience, ceramic products are usually divided into four main types; these are shown below with some examples:

- Structural, including bricks, pipes, floor and roof tiles.
- Refractories, such as kiln linings, gas fire radiants, steel and glass making crucibles.
- Whitewares, including tableware, cookware, wall tiles, pottery products and sanitary ware.
- Technical, also known as engineering, advanced, special, and fine ceramics. Such items include:
    - Gas burner nozzles.
    - Ballistic protection, vehicle armour.
    - Nuclear fuel uranium oxide pellets.
    - Biomedical implants.
    - Coatings of jet engine turbine blades.
    - Ceramic disk brake.
    - Missile nose cones.
    - Bearing (mechanical).
    - Tiles used in the Space Shuttle program.

Ceramics made with Clay

Frequently, the raw materials of modern ceramics do not include clays. Those that do are classified as follows:

- Earthenware, fired at lower temperatures than other types.
- Stoneware, vitreous or semi-vitreous.
- Porcelain, which contains a high content of kaolin.
- Bone china.

Classification of Ceramics

Ceramics can also be classified into three distinct material categories:

- Oxides: Alumina, beryllia, ceria, zirconia.

- Non-oxides: Carbide, boride, nitride, silicide.
- Composite materials: Particulate reinforced, fiber reinforced, combinations of oxides and nonoxides.

Each one of these classes can be developed into unique material properties because ceramics tend to be crystalline.

Applications

- Knife blades: The blade of a ceramic knife will stay sharp for much longer than that of a steel knife, although it is more brittle and susceptible to breaking.
- Carbon-ceramic brake disks for vehicles are resistant to brake fade at high temperatures.
- Advanced composite ceramic and metal matrices have been designed for most modern armoured fighting vehicles because they offer superior penetrating resistance against shaped charges (such as HEAT rounds) and kinetic energy penetrators.
- Ceramics such as alumina and boron carbide have been used in ballistic armored vests to repel high-velocity rifle fire. Such plates are known commonly as small arms protective inserts, or SAPIs. Similar material is used to protect the cockpits of some military airplanes, because of the low weight of the material.
- Ceramics can be used in place of steel for ball bearings. Their higher hardness means they are much less susceptible to wear and typically last for triple the lifetime of a steel part. They also deform less under load, meaning they have less contact with the bearing retainer walls and can roll faster. In very high speed applications, heat from friction during rolling can cause problems for metal bearings, which are reduced by the use of ceramics. Ceramics are also more chemically resistant and can be used in wet environments where steel bearings would rust. In some cases, their electricity-insulating properties may also be valuable in bearings. Two drawbacks to ceramic bearings are a significantly higher cost and susceptibility to damage under shock loads.
- In the early 1980s, Toyota researched production of an adiabatic engine using ceramic components in the hot gas area. The ceramics would have allowed temperatures of over 3000 °F (1650 °C). The expected advantages would have been lighter materials and a smaller cooling system (or no need for one at all), leading to a major weight reduction. The expected increase of fuel efficiency of the engine (caused by the higher temperature, as shown by Carnot's theorem) could not be verified experimentally; it was found that the heat transfer on the hot ceramic cylinder walls was higher than the transfer to a cooler metal wall as the

cooler gas film on the metal surface works as a thermal insulator. Thus, despite all of these desirable properties, such engines have not succeeded in production because of costs for the ceramic components and the limited advantages. (Small imperfections in the ceramic material with its low fracture toughness lead to cracks, which can lead to potentially dangerous equipment failure.) Such engines are possible in laboratory settings, but mass production is not feasible with current technology.

- Work is being done in developing ceramic parts for gas turbine engines. Currently, even blades made of advanced metal alloys used in the engines' hot section require cooling and careful limiting of operating temperatures. Turbine engines made with ceramics could operate more efficiently, giving aircraft greater range and payload for a set amount of fuel.

- Recent advances have been made in ceramics which include bioceramics, such as dental implants and synthetic bones. Hydroxyapatite, the natural mineral component of bone, has been made synthetically from a number of biological and chemical sources and can be formed into ceramic materials. Orthopedic implants coated with these materials bond readily to bone and other tissues in the body without rejection or inflammatory reactions so are of great interest for gene delivery and tissue engineering scaffolds. Most hydroxyapatite ceramics are very porous and lack mechanical strength, and are used to coat metal orthopedic devices to aid in forming a bond to bone or as bone fillers. They are also used as fillers for orthopedic plastic screws to aid in reducing the inflammation and increase absorption of these plastic materials. Work is being done to make strong, fully dense nanocrystalline hydroxyapatite ceramic materials for orthopedic weight bearing devices, replacing foreign metal and plastic orthopedic materials with a synthetic, but naturally occurring, bone mineral. Ultimately, these ceramic materials may be used as bone replacements or with the incorporation of protein collagens, synthetic bones.

- Durable actinide-containing ceramic materials have many applications such as in nuclear fuels for burning excess Pu and in chemically-inert sources of alpha irradiation for power supply of unmanned space vehicles or to produce electricity for microelectronic devices. Both use and disposal of radioactive actinides require their immobilisation in a durable host material. Nuclear waste long-lived radionuclides such as actinides are immobilised using chemically-durable crystalline materials based on polycrystalline ceramics and large single crystals.

- High-tech ceramic is used in watchmaking for producing watch cases. The material is valued by watchmakers for its light weight, scratch resistance, durability and smooth touch. IWC is one of the brands that initiated the use of ceramic in watchmaking.

Archaeology

Ceramic artifacts have an important role in archaeology for understanding the culture, technology and behavior of peoples of the past. They are among the most common artifacts to be found at an archaeological site, generally in the form of small fragments of broken pottery called sherds. Processing of collected sherds can be consistent with two main types of analysis: technical and traditional.

Traditional analysis involves sorting ceramic artifacts, sherds and larger fragments into specific types based on style, composition, manufacturing and morphology. By creating these typologies it is possible to distinguish between different cultural styles, the purpose of the ceramic and technological state of the people among other conclusions. In addition, by looking at stylistic changes of ceramics over time is it possible to separate (seriate) the ceramics into distinct diagnostic groups (assemblages). A comparison of ceramic artifacts with known dated assemblages allows for a chronological assignment of these pieces.

The technical approach to ceramic analysis involves a finer examination of the composition of ceramic artifacts and sherds to determine the source of the material and through this the possible manufacturing site. Key criteria are the composition of the clay and the temper used in the manufacture of the article under study: temper is a material added to the clay during the initial production stage, and it is used to aid the subsequent drying process. Types of temper include shell pieces, granite fragments and ground sherd pieces called 'grog'. Temper is usually identified by microscopic examination of the temper material. Clay identification is determined by a process of refiring the ceramic, and assigning a color to it using Munsell Soil Color notation. By estimating both the clay and temper compositions, and locating a region where both are known to occur, an assignment of the material source can be made. From the source assignment of the artifact further investigations can be made into the site of manufacture.

Glass Ceramics

Glass-ceramics have an amorphous phase and one or more crystalline phases and are produced by a so-called "controlled crystallization" in contrast to a spontaneous crystallization, which is usually not wanted in glass manufacturing. Glass-ceramics have the fabrication advantage of glass, as well as special properties of ceramics. When used for sealing, some glass-ceramics do not require brazing but can withstand brazing temperatures up to 700 °C. Glass-ceramics usually have between 30% [m/m] and 90% [m/m] crystallinity and yield an array of materials with interesting properties like zero porosity, high strength, toughness, translucency or opacity, pigmentation, opalescence, low or even negative thermal expansion, high temperature stability, fluorescence, machinability, ferromagnetism, resorbability or high chemical durability, biocompatibility, bioactivity, ion conductivity, superconductivity, isolation capabilities, low dielectric constant and loss, high resistivity and break-down voltage. These properties can be tailored by controlling the base-glass composition and by controlled heat treatment/crystallization of base glass. In manufacturing, glass-ceramics are valued for having the strength of ceramic but the hermetic sealing properties of glass.

Glass-ceramics are mostly produced in two steps: First, a glass is formed by a glass-manufacturing process. The glass is cooled down and is then reheated in a second step. In this heat treatment the glass partly crystallizes. In most cases nucleation agents are added to the base composition of the glass-ceramic. These nucleation agents aid and control the crystallization process. Because there is usually no pressing and sintering, glass-ceramics have, unlike sintered ceramics, no pores.

A wide variety of glass-ceramic systems exists, e.g., the $Li_2O \times Al_2O_3 \times nSiO_2$ system (LAS system), the $MgO \times Al_2O_3 \times nSiO_2$ system (MAS system), the $ZnO \times Al_2O_3 \times nSiO_2$ system (ZAS system).

The LAS system mainly refers to a mix of lithium, silicon, and aluminum oxides with additional components, e.g., glass-phase-forming agents such as Na_2O, K_2O and CaO and refining agents. As nucleation agents most commonly zirconium(IV) oxide in combination with titanium(IV) oxide is used. This important system was studied first and intensively by Hummel, and Smoke.

After crystallization the dominant crystal phase in this type of glass-ceramic is a high-quartz solid solution (HQ s.s.). If the glass-ceramic is subjected to a more intense heat treatment, this HQ s.s. transforms into a keatite-solid solution (K s.s., sometimes wrongly named as beta-spodumene). This transition is non-reversible and reconstructive, which means bonds in the crystal-lattice are broken and new arranged. However, these two crystal phases show a very similar structure as Li could show.

The most interesting properties of these glass-ceramics are their thermomechanical properties. Glass-ceramic from the LAS system is a mechanically strong material and can sustain repeated and quick temperature changes up to 800–1000 °C. The dominant crystalline phase of the LAS glass-ceramics, HQ s.s., has a strong *negative* coefficient of thermal expansion (CTE), keatite-solid solution as still a negative CTE but much higher than HQ s.s. These negative CTEs of the crystalline phase contrasts with the positive CTE of the residual glass. Adjusting the proportion of these phases offers a wide range of possible CTEs in the finished composite. Mostly for today's applications a low or even zero CTE is desired. Also a negative CTE is possible, which means, in contrast to most materials when heated up, such a glass-ceramic contracts. At a certain point, generally between 60% [m/m] and 80% [m/m] crystallinity, the two coefficients balance such that the glass-ceramic as a whole has a thermal expansion coefficient that is very close to zero. Also, when an interface between material will be subject to thermal fatigue, glass-ceramics can be adjusted to match the coefficient of the material they will be bonded to.

Originally developed for use in the mirrors and mirror mounts of astronomical telescopes, LAS glass-ceramics have become known and entered the domestic market through its use in glass-ceramic cooktops, as well as cookware and bakeware or as high-performance reflectors for digital projectors.

Ceramic Matrix Composites

One particularly notable use of glass-ceramics is in the processing of ceramic matrix composites. For many ceramic matrix composites typical sintering temperatures and times cannot be used, as the degradation and corrosion of the constituent fibres becomes more of an issue as temperature and sintering time increase. One example of this is SiC fibres, which can start to degrade via pyrolysis at temperatures above 1470K. One solution to this is to use the glassy form of the ceramic as the sintering feedstock rather than the ceramic, as unlike the ceramic the glass pellets have a softening point and will generally flow at much lower pressures and temperatures. This allows the use of less extreme processing parameters, making the production of many new technologically important fibre-matrix combinations by sintering possible.

Cooktops

Glass-ceramic from the LAS-System is a mechanically strong material and can sustain repeated and quick temperature changes. However, it is not totally unbreakable. Because it is still a brittle material as glass and ceramics are, it can be broken. There have been instances where users reported damage to their cooktops when the surface was struck with a hard or blunt object (such as a can falling from above or other heavy items).

The material has a very low heat conduction coefficient, which means that it stays cool outside the cooking area. It can be made nearly transparent (15–20% loss in a typical cooktop) for radiation in the infrared wavelengths.

In the visible range glass-ceramics can be transparent, translucent or opaque and even colored by coloring agents.

A glass-ceramic cooktop.

Today, there are two major types of electrical stoves with cooktops made of glass-ceramic:

- A glass-ceramic stove uses radiant heating coils or infrared halogen lamps as the heating elements. The surface of the glass-ceramic cooktop above the burner heats up, but the adjacent surface remains cool because of the low heat conduction coefficient of the material.
- An induction stove heats a metal pot's bottom directly through electromagnetic induction.

This technology is not entirely new, as glass-ceramic ranges were first introduced in the 1970s using Corningware tops instead of the more durable material used today. These first generation smoothtops were problematic and could only be used with flat-bottomed cookware as the heating was primarily conductive rather than radiative.

Compared to conventional kitchen stoves, glass-ceramic cooktops are relatively simple to clean, due to their flat surface. However, glass-ceramic cooktops can be scratched very easily, so care must be taken not to slide the cooking pans over the surface. If food with a high sugar content (such as jam) spills, it should never be allowed to dry on the surface, otherwise damage will occur.

For best results and maximum heat transfer, all cookware should be flat-bottomed and matched to the same size as the burner zone.

Brands and Manufacturers

CorningWare casserole dish and other cookware pieces, with the 'Cornflower' pattern decoration.

Some well-known brands of glass-ceramics are Pyroceram, *Ceran* (cooktops), *Eurokera* (cooktop, stoves and fireplaces), *Zerodur* (telescope mirrors), and *Macor*, *Kanger* (Glass-Ceramic for cooktop, stoves, microwave and fireplaces door). German manufacturer Schott introduced Zerodur in 1968, Ceran followed in 1971.

Nippon Electric Glass is a predominent worldwide manufacturer of glass ceramics, whose related products in this area include FireLite and NeoCeram ceramic glass for architectural and high temperature applications respectively. Keralite, manufactured by Vetrotech Saint-Gobain, is a specialty glass-ceramic fire and impact safety rated material for use in fire-rated applications. Glass-ceramics manufactured in the Soviet Union/Russia are known under the name *Sitall*.

The same class of material was also used, until the late 1990s, in CorningWare dishes, which could be taken from the freezer directly to the oven with no risk of thermal shock.

Crystal Structure of Ceramics

Ceramic Glass

Ceramics with an entirely glassy structure have certain properties that are quite different from those of metals. Recall that when metal in the liquid state is cooled, a crystalline solid precipitates when the melting freezing point is reached. However, with a glassy material, as the liquid is cooled it becomes more and more viscous. There is no sharp melting or freezing point. It goes from liquid to a soft plastic solid and finally becomes hard and brittle. Because of this unique property, it can be blown into shapes, in addition to being cast, rolled, drawn and otherwise processed like a metal.

Glassy behavior is related to the atomic structure of the material. If pure silica (SiO_2) is fused together, a glass called vitreous silica is formed on cooling. The basic unit structure of this glass is the silica tetrahedron, which is composed of a single silicon atom surrounded by four equidistant oxygen atoms. The silicon atoms occupy the openings (interstitials) between the oxygen atoms and share four valence electrons with the oxygen atoms through covalent bonding. The silica atom has four valence electrons and each of the oxygen atoms has two valence electrons so the silica tetrahedron has four extra valence electrons to share with adjacent tetrahedral. The silicate structures can link together by sharing the atoms in two corners of the SiO_2 tetrahedrons, forming chain or ring structures. A network of silica tetrahedral chains form, and at high temperatures these chains easily slide past each other. As the melt cools, thermal vibrational energy decreases and the chains can not move as easily so the structure becomes more rigid. Silica is the most important constituent of glass, but other oxides are added to change certain physical characteristics or to lower the melting point.

Ceramic Crystalline or Partially Crystalline Material

Most ceramics usually contain both metallic and nonmetallic elements with ionic or covalent bonds. Therefore, the structure the metallic atoms, the structure of the nonmetallic atoms, and the balance of charges produced by the valence electrons must be considered. As with metals, the unit cell is used in describing the atomic structure of ceramics. The cubic and the hexagonal cells are most common. Additionally, the difference in radii between the metallic and nonmetallic ions plays an important role in the arrangement of the unit cell.

In metals, the regular arrangement of atoms into densely packed planes led to the occurrence of slip under stress, which gives metal their characteristic ductility. In ceramics, brittle fracture rather than slip is common because both the arrangement of the atoms and the type of bonding is different. The fracture or cleavage planes of ceramics are the result of planes of regularly arranged atoms.

The building criteria for the crystal structure are:

- Maintain neutrality.
- Charge balance dictates chemical formula.
- Achieve closest packing.

Silicate Ceramics

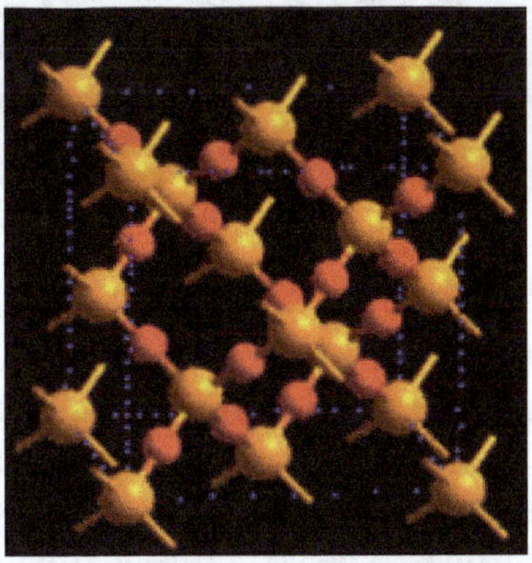

The silica structure is the basic structure for many ceramics, as well as glass. It has an internal arrangement consisting of pyramid (tetrahedral or four-sided) units. Four large oxygen (o) atoms surround each smaller silicon (Si) atom. When silica tetrahedrons share three corner atoms, they produce layered silicates (talc, kaolinite clay, mica). Clay is the basic raw material for many building products such as brick and tile. When silica tetrahedrons share four comer atoms, they produce framework silicates (quartz, tridymite). Quartz is formed when the tetrahedra in this material are arranged in a regular, orderly fashion. If silica in the molten state is cooled very slowly it crystallizes at the freezing point. But if molten silica is cooled more rapidly, the resulting solid is a disorderly arrangement which is glass.

Cement

Cement (Portland cement) is one of the main ingredients of concrete. There are a number of different grades of cement but a typical Portland cement will contain 19 to 25% SiO_2, 5 to 9% Al_2O_3, 60 to 64% CaO and 2 to 4% FeO. Cements are prepared by grinding the clays and limestone in proper proportion, firing in a kiln, and regrinding. When water is added, the minerals either decompose or combine with water, and a new phase grows throughout the mass. The reaction is solution, recrystallization, and precipitation of a silicate structure. It is usually important to control the amount of water to prevent an excess that would not be part of the structure and would weaken it. The heat of hydration (heat of reaction in the adsorption of water) in setting of the cement can be large and can cause damage in large structures.

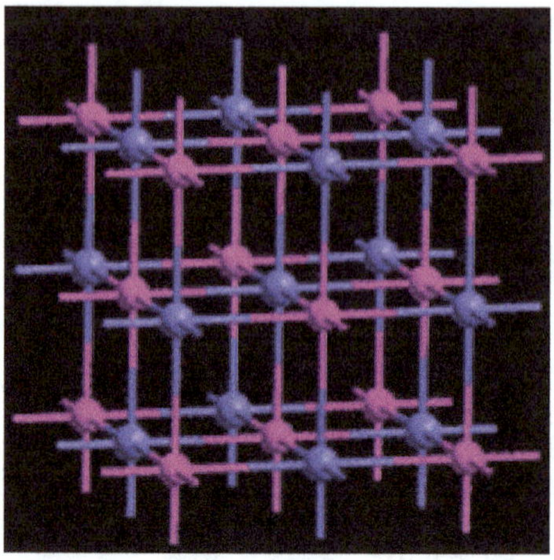

Nitride Ceramics

Nitrides combine the superior hardness of ceramics with high thermal and mechanical stability, making them suitable for applications as cutting tools, wear-resistant parts and structural components at high temperatures. TiN has a cubic structure which is perhaps the simplest and best known of structure types. Cations and anions both lie at the nodes of separate fcc lattices. The structure is unchanged if the Ti and N atoms (lattices) are interchanged.

Ferroelectric Ceramics

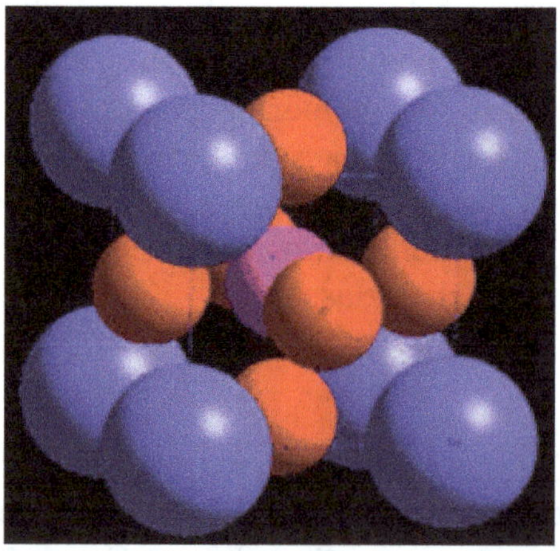

Depending on the crystal structure, in some crystal lattices, the centers of the positive and negative charges do not coincide even without the application of external electric field. In this case, it is said that there exists spontaneous polarization in the crystal. When the polarization of the dielectric can be altered by an electric field, it is called ferroelectric. A typical ceramic ferroelectric is barium titanate, $BaTiO_3$. Ferroelectric materials, especially polycrystalline ceramics, are very promising for varieties of application fields such as piezoelectric/electrostrictive transducers, and electrooptic.

Phase Diagram

The phase diagram is important in understanding the formation and control of the microstructure of the microstructure of polyphase ceramics, just as it is with polyphase metallic materials. Also, nonequilibrium structures are even more prevalent in ceramics because the more complex crystal structures are more difficult to nucleate and to grow from the melt.

Imperfections in Ceramics

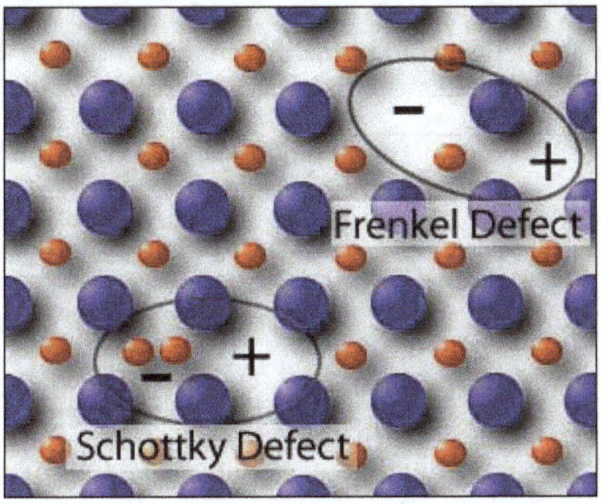

Imperfections in ceramic crystals include point defects and impurities like in metals. However, in ceramics defect formation is strongly affected by the condition of charge neutrality because the creation of areas of unbalanced charges requires an expenditure of a large amount of energy. In ionic crystals, charge neutrality often results in defects that come as pairs of ions with opposite charge or several nearby point defects in which the sum of all charges is zero. Charge neutral defects include the Frenkel and Schottky defects. A Frenkel-defect occurs when a host atom moves into a nearby interstitial position to create a vacancy-interstitial pair of cations. A Schottky-defect is a pair of nearby cation and anion vacancies. Schottky defect occurs when a host atom leaves its position and moves to the surface creating a vacancy-vacancy pair.

Sometimes, the composition may alter slightly to arrive at a more balanced atomic charge. Solids such as SiO_2, which have a well-defined chemical formula, are called stoichiometric compounds. When the composition of a solid deviates from the standard chemical formula, the resulting solid is said to be nonstoichiometric. Nonstoichiometry and the existence of point defects in a solid are often closely related. Anion vacancies are the source of the nonstoichiometry in SiO_{2-x}.

Introduction of impurity atoms in the lattice is likely in conditions where the charge is maintained. This is the case of electronegative impurities that substitute a lattice anion or electropositive substitutional impurities. This is more likely for similar ionic radii since this minimizes the energy required for lattice distortion. Defects will appear if the charge of the impurities is not balanced.

Composite Materials

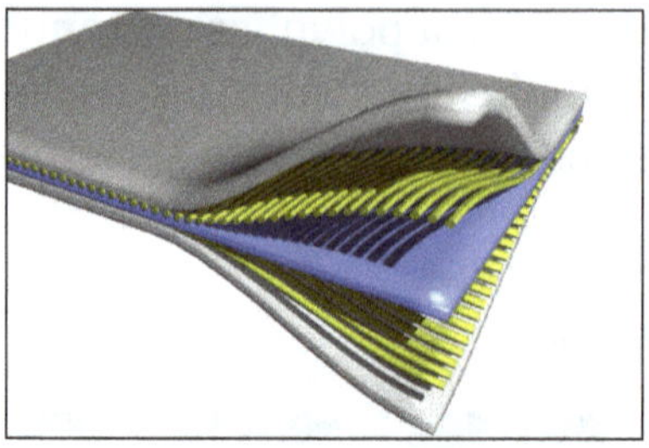

Composites are formed by combining materials together to form an overall structure with properties that differ from that of the individual components.

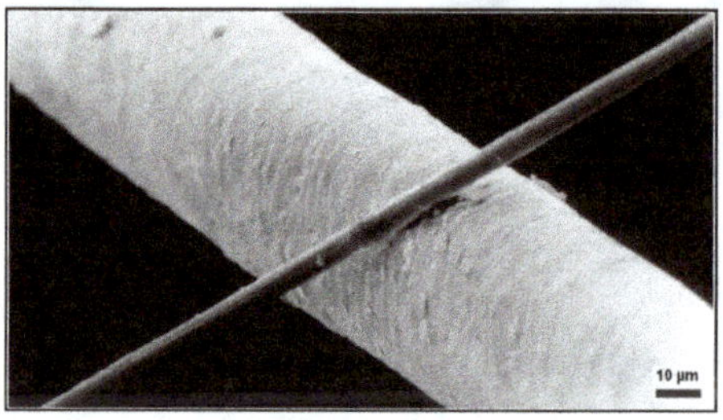

A black carbon fibre (used as a reinforcement component) compared to a human hair.

A composite material (also called a composition material or shortened to composite, which is the common name) is a material made from two or more constituent materials with significantly different physical or chemical properties that, when combined, produce a material with characteristics different from the individual components. The individual components remain separate and distinct within the finished structure, differentiating composites from mixtures and solid solutions.

The new material may be preferred for many reasons. Common examples include materials which are stronger, lighter, or less expensive when compared to traditional materials.

More recently, researchers have also begun to actively include sensing, actuation, computation and communication into composites, which are known as Robotic Materials.

Typical engineered composite materials include:

- Reinforced concrete and masonry.
- Composite wood such as plywood.
- Reinforced plastics, such as fibre-reinforced polymer or fiberglass.
- Ceramic matrix composites (composite ceramic and metal matrices).
- Metal matrix composites.
- And other Advanced composite materials.

Composite materials are generally used for buildings, bridges, and structures such as boat hulls, swimming pool panels, racing car bodies, shower stalls, bathtubs, storage tanks, imitation granite and cultured marble sinks and countertops.

The most advanced examples perform routinely on spacecraft and aircraft in demanding environments.

Examples:

Concrete is the most common artificial composite material of all and typically consists of loose stones (aggregate) held with a matrix of cement. Concrete is an inexpensive material, and will not compress or shatter even under quite a large compressive force. However, concrete cannot survive tensile loading(i.e., if stretched it will quickly break apart). Therefore, to give concrete the ability to resist being stretched, steel bars, which can resist high stretching forces, are often added to concrete to form reinforced concrete.

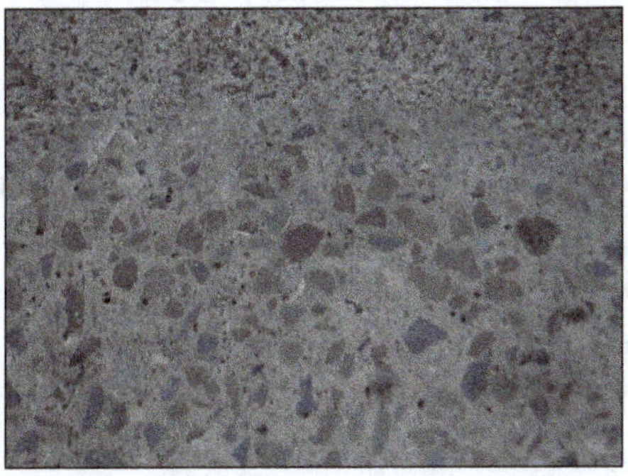

Concrete is a mixture of cement and aggregate, giving a robust, strong material that is very widely used.

Fibre-reinforced polymers (FRP)s include carbon-fibre-reinforced polymer (CFRP) and glass-reinforced plastic (GRP). If classified by matrix then there are thermoplastic composites, short fibre thermoplastics, long fibre thermoplastics or long fibre-reinforced thermoplastics. There are numerous thermoset composites, including paper composite panels. Many advanced thermoset polymer matrix systems usually incorporate aramid fibre and carbon fibre in an epoxy resin matrix.

Plywood is used widely in construction.

Composite sandwich structure panel used for testing at NASA.

Shape memory polymer composites are high-performance composites, formulated using fibre or fabric reinforcement and shape memory polymer resin as the matrix. Since a shape memory polymer resin is used as the matrix, these composites have the ability to be easily manipulated into various configurations when they are heated above their activation temperatures and will exhibit high strength and stiffness at lower temperatures. They can also be reheated and reshaped repeatedly without losing their material properties. These composites are ideal for applications such as lightweight, rigid, deployable structures; rapid manufacturing; and dynamic reinforcement.

High strain composites are another type of high-performance composites that are designed to perform in a high deformation setting and are often used in deployable systems where structural flexing is advantageous. Although high strain composites exhibit many similarities to shape memory polymers, their performance is generally dependent on the fibre layout as opposed to the resin content of the matrix.

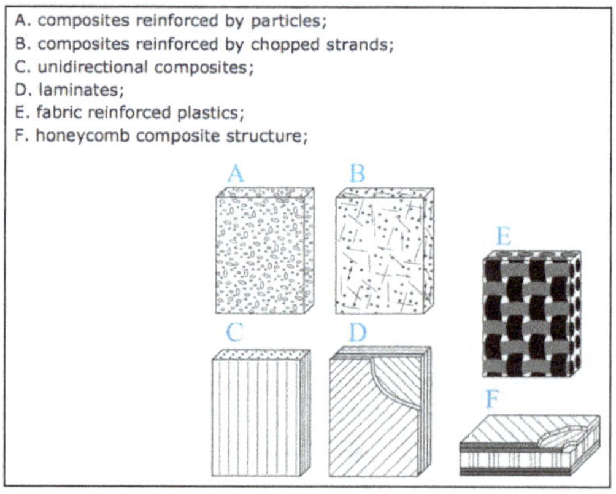

"Structural Integrity Analysis: Composites".

Composites can also use metal fibres reinforcing other metals, as in metal matrix composites (MMC) or ceramic matrix composites (CMC), which includes bone (hydroxyapatite reinforced with collagen fibres), cermet (ceramic and metal) and concrete. Ceramic matrix composites are built primarily for fracture toughness, not for strength. Another class of composite materials involve woven fabric composite consisting of longitudinal and transverse laced yarns. Woven fabric composites are flexible as they are in form of fabric.

Organic matrix/ceramic aggregate composites include asphalt concrete, polymer concrete, mastic asphalt, mastic roller hybrid, dental composite, syntactic foam and mother of pearl. Chobham armour is a special type of composite armour used in military applications.

Additionally, thermoplastic composite materials can be formulated with specific metal powders resulting in materials with a density range from 2 g/cm^3 to 11 g/cm^3 (same density as lead). The most common name for this type of material is "high gravity compound" (HGC), although "lead replacement" is also used. These materials can be used in place of traditional materials such as aluminium, stainless steel, brass, bronze, copper, lead, and even tungsten in weighting, balancing (for example, modifying the centre of gravity of a tennis racquet), vibration damping, and radiation shielding applications. High density composites are an economically viable option when certain materials are deemed hazardous and are banned (such as lead) or when secondary operations costs (such as machining, finishing, or coating) are a factor.

A sandwich-structured composite is a special class of composite material that is fabricated by attaching two thin but stiff skins to a lightweight but thick core. The core material is normally low strength material, but its higher thickness provides the sandwich composite with high bending stiffness with overall low density.

Wood is a naturally occurring composite comprising cellulose fibres in a lignin and hemicellulose matrix. Engineered wood includes a wide variety of different products such as wood fibre board, plywood, oriented strand board, wood plastic composite (recycled wood fibre in polyethylene matrix), Pykrete (sawdust in ice matrix), Plastic-impregnated or laminated paper or textiles, Arborite, Formica (plastic) and Micarta. Other engineered laminate composites, such as Mallite, use a central core of end grain balsa wood, bonded to surface skins of light alloy or GRP. These generate low-weight, high rigidity materials.

Particulate composites have particle as filler material dispersed in matrix, which may be nonmetal, such as glass, epoxy. Automobile tire is an example of particulate composite.

Advanced diamond-like carbon (DLC) coated polymer composites have been reported where the coating increases the surface hydrophobicity, hardness and wear resistance.

Products

Fibre-reinforced composite materials have gained popularity (despite their generally high cost) in high-performance products that need to be lightweight, yet strong enough to take harsh loading conditions such as aerospace components (tails, wings, fuselages, propellers), boat and scull hulls, bicycle frames and racing car bodies. Other uses include fishing rods, storage tanks, swimming pool panels, and baseball bats. The Boeing 787 and Airbus A350 structures including the wings and fuselage are composed largely of composites. Composite materials are also becoming more common in the realm of orthopedic surgery, and it is the most common hockey stick material.

Carbon composite is a key material in today's launch vehicles and heat shields for the re-entry phase of spacecraft. It is widely used in solar panel substrates, antenna reflectors and yokes of spacecraft. It is also used in payload adapters, inter-stage structures and heat shields of launch vehicles. Furthermore, disk brake systems of airplanes and racing cars are using carbon/carbon material, and the composite material with carbon fibres and silicon carbide matrix has been introduced in luxury vehicles and sports cars.

In 2006, a fibre-reinforced composite pool panel was introduced for in-ground swimming pools, residential as well as commercial, as a non-corrosive alternative to galvanized steel.

In 2007, an all-composite military Humvee was introduced by TPI Composites Inc and Armor Holdings Inc, the first all-composite military vehicle. By using composites the vehicle is lighter, allowing higher payloads. In 2008, carbon fibre and DuPont Kevlar (five times stronger than steel) were combined with enhanced thermoset resins to make military transit cases by ECS Composites creating 30-percent lighter cases with high strength.

Pipes and fittings for various purpose like transportation of potable water, fire-fighting, irrigation, seawater, desalinated water, chemical and industrial waste, and sewage are now manufactured in glass reinforced plastics.

Composite materials used in tensile structures for facade application provides the advantage of being translucent. The woven base cloth combined with the appropriate coating allows better light transmission. This provides a very comfortable level of illumination compared to the full brightness of outside.

The wings of wind turbines, in growing sizes in the order of 50 m length are fabricated in composites since several years.

Two-lower-leg-amputees run on carbon-composite spring-like artificial feet as quick as healthy sportsmen.

High pressure gas cylinders typically about 7–9 litre volume x 300 bar pressure for firemen are nowadays constructed from carbon composite. Type-4-cylinders include metal only as boss that carries the thread to screw in the valve.

On 5 September 2019, HMD Global unveiled the Nokia 6.2 and Nokia 7.2 which are claimed to be using polymer composite for the frames.

Composites are made up of individual materials referred to as constituent materials. There are two main categories of constituent materials: matrix (binder) and reinforcement. At least one portion of each type is required. The matrix material surrounds and supports the reinforcement materials by maintaining their relative positions. The reinforcements impart their special mechanical and physical properties to enhance the matrix properties. A synergism produces material properties unavailable from the individual constituent materials, while the wide variety of matrix and strengthening materials allows the designer of the product or structure to choose an optimum combination.

Carbon fibre composite part.

Engineered composite materials must be formed to shape. The matrix material can be introduced to the reinforcement before or after the reinforcement material is placed into the mould cavity or onto the mould surface. The matrix material experiences a melding event, after which the part shape is essentially set. Depending upon the nature of the matrix material, this melding event can occur in various ways such as chemical polymerization for a thermoset polymer matrix, or solidification from the melted state for a thermoplastic polymer matrix composite.

A variety of moulding methods can be used according to the end-item design requirements. The principal factors impacting the methodology are the natures of the chosen matrix and reinforcement materials. Another important factor is the gross quantity of material to be produced. Large quantities can be used to justify high capital expenditures for rapid and automated manufacturing technology. Small production quantities are accommodated with lower capital expenditures but higher labour and tooling costs at a correspondingly slower rate.

Many commercially produced composites use a polymer matrix material often called a resin solution. There are many different polymers available depending upon the starting raw ingredients. There are several broad categories, each with numerous variations. The most common are known as polyester, vinyl ester, epoxy, phenolic, polyimide, polyamide, polypropylene, PEEK, and others. The reinforcement materials are often fibres but also commonly ground minerals. The various methods described below have been developed to reduce the resin content of the final product, or the fibre content is increased. As a rule of thumb, lay up results in a product containing 60% resin and 40% fibre, whereas vacuum infusion gives a final product with 40% resin and 60% fibre content. The strength of the product is greatly dependent on this ratio.

Martin Hubbe and Lucian A Lucia consider wood to be a natural composite of cellulose fibres in a matrix of lignin.

Constituents

Matrices

Polymers are common matrices (especially used for fibre reinforced plastics). Road surfaces are often made from asphalt concrete which uses bitumen as a matrix. Mud (wattle and daub) has seen extensive use. Typically, most common polymer-based composite materials, including fibreglass, carbon fibre, and Kevlar, include at least two parts, the substrate and the resin.

Polyester resin tends to have yellowish tint, and is suitable for most backyard projects. Its weaknesses are that it is UV sensitive and can tend to degrade over time, and thus generally is also coated to help preserve it. It is often used in the making of surfboards and for marine applications. Its hardener is a peroxide, often MEKP (methyl ethyl ketone peroxide). When the peroxide is mixed with the resin, it decomposes to generate free radicals, which initiate the curing reaction. Hardeners in these systems are commonly called catalysts, but since they do not re-appear unchanged at the end of the reaction, they do not fit the strictest chemical definition of a catalyst.

Vinyl ester resin tends to have a purplish to bluish to greenish tint. This resin has lower viscosity than polyester resin and is more transparent. This resin is often billed as being fuel resistant, but will melt in contact with gasoline. It tends to be more resistant over time to degradation than polyester resin and is more flexible. It uses the same hardeners as polyester resin (at a similar mix ratio) and the cost is approximately the same.

Epoxy resin is almost transparent when cured. In the aerospace industry, epoxy is used as a structural matrix material or as a structural glue.

Shape memory polymer (SMP) resins have varying visual characteristics depending on their formulation. These resins may be epoxy-based, which can be used for auto body and outdoor equipment repairs; cyanate-ester-based, which are used in space applications; and acrylate-based, which can be used in very cold temperature applications, such as for sensors that indicate whether perishable goods have warmed above a certain maximum temperature. These resins are unique in that their shape can be repeatedly changed by heating above their glass transition temperature (T_g). When heated, they become flexible and elastic, allowing for easy configuration. Once they are cooled, they will maintain their new shape. The resins will return to their original shapes when they are reheated above their T_g. The advantage of shape memory polymer resins is that they can be shaped and reshaped repeatedly without losing their material properties. These resins can be used in fabricating shape memory composites.

Traditional materials such as glues, muds have traditionally been used as matrices for papier-mâché and adobe.

Inorganic

Cement (concrete), metals, ceramics, and sometimes glasses are employed. Unusual matrices such as ice are sometime proposed as in pykecrete.

Reinforcements

Fiber

Differences in the way the fibres are laid out give different strengths and ease of manufacture.

Reinforcement usually adds rigidity and greatly impedes crack propagation. Thin fibers can have very high strength, and provided they are mechanically well attached to the matrix they can greatly improve the composite's overall properties.

Fibre-reinforced composite materials can be divided into two main categories normally referred to as short fibre-reinforced materials and continuous fiber-reinforced materials. Continuous reinforced materials will often constitute a layered or laminated structure. The woven and continuous fiber styles are typically available in a variety of forms, being pre-impregnated with the given matrix (resin), dry, uni-directional tapes of various widths, plain weave, harness satins, braided, and stitched.

The short and long fibres are typically employed in compression moulding and sheet moulding operations. These come in the form of flakes, chips, and random mate (which can also be made from a continuous fibre laid in random fashion until the desired thickness of the ply / laminate is achieved).

Common fibres used for reinforcement include glass fibres, carbon fibres, cellulose (wood/paper fibre and straw) and high strength polymers for example aramid. Silicon carbide fibers are used for some high temperature applications.

Particle

Particle reinforcement adds a similar effect to precipitation hardening in metals and ceramics. Large particles impede dislocation movement and crack propagation as well as contribute to the composite's Young's Modulus. In general, particle reinforcement effect on Young's Modulus lies between values predicted by:

$$E_c = \frac{E_\alpha E_\beta}{(V_\alpha E_\beta + V_\beta E_\alpha)}$$

as a lower bound and

$$E_c = V_\alpha E_\alpha + V_\beta E_\beta$$

as an upper bound.

Therefore, it can be expressed as a linear combination of contribution from the matrix and some weighted contribution from the particles.

$$E_c = V_m E_m + K_c V_p E_p$$

Where K_c is an experimentally derived constant between 0 and 1. This range of values for K_c reflects that particle reinforced composites are not characterized by the isostrain condition.

Similarly, the tensile strength can be modeled in an equation of similar construction where K_s is a similarly bounded constant not necessarily of the same value of K_c:

$$(T.S.)_c = V_m (T.S.)_m + K_s V_p (T.S.)_p$$

The true value of K_c and K_s vary based on factors including particle shape, particle distribution, and particle/matrix interface. Knowing these parameters, the mechanical properties can be modeled based on effects from grain boundary strengthening, dislocation strengthening, and Orowan strengthening.

The most common particle reinforced composite is concrete, which is a mixture of gravel and sand usually strengthened by addition of small rocks or sand. Metals are often reinforced with ceramics to increase strength at the cost of ductility. Finally polymers and rubber are often reinforced with carbon black, commonly used in auto tires.

Cores

Many composite layup designs also include a co-curing or post-curing of the prepreg with various other media, such as honeycomb or foam. This is commonly called a sandwich structure. This is a more common layup for the manufacture of radomes, doors, cowlings, or non-structural parts.

Open- and closed-cell-structured foams like polyvinylchloride, polyurethane, polyethylene or polystyrene foams, balsa wood, syntactic foams, and honeycombs are commonly used core materials. Open- and closed-cell metal foam can also be used as core materials. Recently, 3D graphene structures (also called graphene foam) have also been employed as core structures. A recent review by Khurram and Xu et al., have provided the summary of the state-of-the-art techniques for fabrication of the 3D structure of graphene, and the examples of the use of these foam like structures as a core for their respective polymer composites.

Semi-Crystalline Polymers

Although the two phases are chemically equivalent, semi-crystalline polymers can be described both quantitatively and qualitatively as composite materials. The crystalline portion has a higher elastic modulus and provides reinforcement for the less stiff, amorphous phase. Polymeric materials can range from 0% to 100% crystallinity aka volume fraction depending on molecular structure and thermal history. Different processing techniques can be employed to vary the percent crystallinity in these materials and thus the mechanical properties of these materials as described in the physical properties section. This effect is seen in a variety of places from industrial plastics like polyethylene shopping bags to spiders which can produce silks with different mechanical properties. In many cases these materials act like particle composites with randomly dispersed crystals known as spherulites. However they can also be engineered to be anisotropic and act more like fiber reinforced composites. In the case of spider silk, the properties of the material can even be dependent on the size of the crystals, independent of the volume fraction. Ironically, single component polymeric materials are some of the most easily tunable composite materials known.

Fabrication Methods

Fabrication of composite materials is accomplished by a wide variety of techniques, including:

- Advanced fibre placement (Automated fibre placement).
- Tailored fibre placement.
- fibreglass spray lay-up process.
- Filament winding.
- Lanxide process.
- Tufting.
- Z-pinning.

Composite fabrication usually involves wetting, mixing or saturating the reinforcement with the matrix, and then causing the matrix to bind together (with heat or a chemical reaction) into a rigid structure. The operation is usually done in an open or closed forming mould, but the order and ways of introducing the ingredients varies considerably.

Mould Overview

Within a mould, the reinforcing and matrix materials are combined, compacted, and cured (processed) to undergo a melding event. After the melding event, the part shape is essentially set, although it can deform under certain process conditions. For a thermoset polymer matrix material, the melding event is a curing reaction that is initiated by the application of additional heat or chemical reactivity such as an organic peroxide. For a thermoplastic polymeric matrix material, the melding event is a solidification from the melted state. For a metal matrix material such as titanium foil, the melding event is a fusing at high pressure and a temperature near the melting point.

For many moulding methods, it is convenient to refer to one mould piece as a "lower" mould and another mould piece as an "upper" mould. Lower and upper refer to the different faces of the moulded panel, not the mould's configuration in space. In this convention, there is always a lower mould, and sometimes an upper mould. Part construction begins by applying materials to the lower mould. Lower mould and upper mould are more generalized descriptors than more common and specific terms such as male side, female side, a-side, b-side, tool side, bowl, hat, mandrel, etc. Continuous manufacturing uses a different nomenclature.

The moulded product is often referred to as a panel. For certain geometries and material combinations, it can be referred to as a casting. For certain continuous processes, it can be referred to as a profile.

Vacuum Bag Moulding

Vacuum bag moulding uses a flexible film to enclose the part and seal it from outside air. Vacuum bag material is available in a tube shape or a sheet of material. A vacuum is then drawn on the vacuum bag and atmospheric pressure compresses the part during the cure. When a tube shaped bag is used, the entire part can be enclosed within the bag. When using sheet bagging materials, the edges of the vacuum bag are sealed against the edges of the mould surface to enclose the part against an air-tight mould. When bagged in this way, the lower mold is a rigid structure and the upper surface of the part is formed by the flexible membrane vacuum bag. The flexible membrane can be a reusable silicone material or an extruded polymer film. After sealing the part inside the vacuum bag, a vacuum is drawn on the part (and held) during cure. This process can be performed at either ambient or elevated temperature with ambient atmospheric pressure acting upon the vacuum bag. A vacuum pump is typically used to draw a vacuum. An economical method of drawing a vacuum is with a venturi vacuum and air compressor.

A vacuum bag is a bag made of strong rubber-coated fabric or a polymer film used to compress the part during cure or hardening. In some applications the bag encloses the entire material, or in other applications a mold is used to form one face of the laminate with the bag being a single layer to seal to the outer edge of the mold face. When using a tube shaped bag, the ends of the bag are sealed and the air is drawn out of the bag through a nipple using a vacuum pump. As a result, uniform pressure approaching one atmosphere is applied to the surfaces of the object inside the bag, holding parts together while the adhesive cures. The entire bag may be placed in a temperature-controlled oven, oil bath or water bath and gently heated to accelerate curing.

Vacuum bagging is widely used in the composites industry as well. Carbon fibre fabric and fibreglass, along with resins and epoxies are common materials laminated together with a vacuum bag operation.

Woodworking Applications

In commercial woodworking facilities, vacuum bags are used to laminate curved and irregular shaped workpieces.

Typically, polyurethane or vinyl materials are used to make the bag. A tube shaped bag is open at both ends. The piece, or pieces to be glued are placed into the bag and the ends sealed. One method of sealing the open ends of the bag is by placing a clamp on each end of the bag. A plastic rod is laid across the end of the bag, the bag is then folded over the rod. A plastic sleeve with an opening in it, is then snapped over the rod. This procedure forms a seal at both ends of the bag, when the vacuum is ready to be drawn.

A "platen" is sometimes used inside the bag for the piece being glued to lie on. The platen has a series of small slots cut into it, to allow the air under it to be evacuated. The platen must have rounded edges and corners to prevent the vacuum from tearing the bag.

When a curved part is to be glued in a vacuum bag, it is important that the pieces being glued be placed over a solidly built form, or have an air bladder placed under the form. This air bladder has access to "free air" outside the bag. It is used to create an equal pressure under the form, preventing it from being crushed.

Pressure Bag Moulding

This process is related to vacuum bag molding in exactly the same way as it sounds. A solid female mold is used along with a flexible male mold. The reinforcement is placed inside the female mold with just enough resin to allow the fabric to stick in place (wet lay up). A measured amount of resin is then liberally brushed indiscriminately into the mold and the mold is then clamped to a machine that contains the male flexible mold. The flexible male membrane is then inflated with heated compressed air or possibly steam. The female mold can also be heated. Excess resin is forced out along with trapped air. This process is extensively used in the production of composite helmets due to the lower cost of unskilled labor. Cycle times for a helmet bag moulding machine vary from 20 to 45 minutes, but the finished shells require no further curing if the molds are heated.

Autoclave Moulding

A process using a two-sided mould set that forms both surfaces of the panel. On the lower side is a rigid mould and on the upper side is a flexible membrane made from silicone or an extruded polymer film such as nylon. Reinforcement materials can be placed manually or robotically. They include continuous fibre forms fashioned into textile constructions. Most often, they are pre-impregnated with the resin in the form of prepreg fabrics or unidirectional tapes. In some instances, a resin film is placed upon the lower mould and dry reinforcement is placed above. The upper mould is installed and vacuum is applied to the mould cavity. The assembly is placed into an autoclave. This process is generally performed at both elevated pressure and elevated temperature. The use of elevated pressure facilitates a high fibre volume fraction and low void content for maximum structural efficiency.

Resin Transfer Moulding (RTM)

RTM is a process using a rigid two-sided mould set that forms both surfaces of the panel. The mould is typically constructed from aluminum or steel, but composite molds are sometimes used. The two sides fit together to produce a mould cavity. The distinguishing feature of resin transfer moulding is that the reinforcement materials are placed into this cavity and the mould set is closed prior to the introduction of matrix material. Resin transfer moulding includes numerous varieties which differ in the mechanics of how the resin is introduced to the reinforcement in the mould cavity. These variations include everything from the RTM methods used in out of autoclave composite manufacturing for high-tech aerospace components to vacuum infusion to vacuum assisted resin transfer moulding (VARTM). This process can be performed at either ambient or elevated temperature and is suitable for manufacturing high performance composite components in medium volumes (1,000s to 10,000s of parts).

Light Resin Transfer Molding (LRTM)

Similar to the methods performed in Resin Transfer Molding, Light Resin Transfer Molding (Light RTM) involves a closed mold process. A vacuum holds mold A and mold B together to result in two finished sides with fixed thickness levels. Vacuum rings around the tools hold the molds together for this process after dry fiber reinforcements are loaded into mold A before joining with mold B. The air is vacuumed out of the molds with a lower vacuum level, separate from the tooling. After the air is removed the resin is injected into the part. The vacuum remains in effect into the resin is cured.

Other Fabrication Methods

Other types of fabrication include press moulding, transfer moulding, pultrusion moulding, filament winding, casting, centrifugal casting, braiding (onto a former), continuous casting and slip forming. There are also forming capabilities including CNC filament winding, vacuum infusion, wet lay-up, compression moulding, and thermoplastic moulding, to name a few. The use of curing ovens and paint booths is also needed for some projects.

Finishing Methods

The finishing of the composite parts is also critical in the final design. Many of these finishes will include rain-erosion coatings or polyurethane coatings.

Tooling

The mould and mould inserts are referred to as "tooling." The mould/tooling can be constructed from a variety of materials. Tooling materials include invar, steel, aluminium, reinforced silicone rubber, nickel, and carbon fibre. Selection of the tooling material is typically based on, but not limited to, the coefficient of thermal expansion, expected number of cycles, end item tolerance, desired or required surface condition, method of cure, glass transition temperature of the material being moulded, moulding method, matrix, cost and a variety of other considerations.

Physical Properties

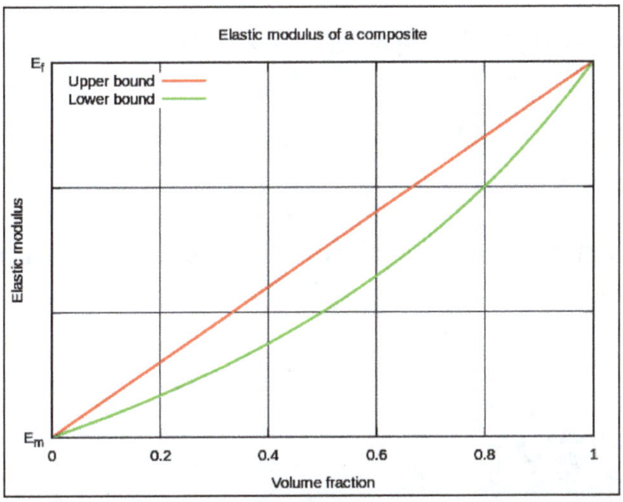

Plot of the overall strength of a composite material as a function of fiber volume fraction limited by the upper bound (isostrain) and lower bound (isostress) conditions.

The physical properties of composite materials are generally not isotropic (independent of direction of applied force) in nature, but they are typically anisotropic (different depending on the direction of the applied force or load). For instance, the stiffness of a composite panel will often depend upon the orientation of the applied forces and/or moments. The strength of a composite is bounded by two loading conditions as shown in the plot to the right. If both the fibres and matrix are aligned parallel to the loading direction, the deformation of both phases will be the same (assuming there is no delamination at the fibre-matrix interface). This isostrain condition provides the upper bound for composite strength, and is determined by the rule of mixtures:

$$E_C = \sum_{i=1} V_i E_i$$

where E_C is the effective composite Young's modulus, and V_i and E_i are the volume fraction and Young's moduli, respectively, of the composite phases.

For example, a composite material made up of α and β phases as shown in the figure to the right under isostrain, the Young's modulus would be as follows:

$$E_C = V_\alpha E_\alpha + V_\beta E_\beta$$

where V_α and V_β are the respective volume fractions of each phase. This can be derived by considering that in the isostrain case,

$$\epsilon_C = \epsilon_\alpha = \epsilon_\beta = \epsilon$$

Assuming that the composite has a uniform cross section, the stress on the composite is a weighted average between the two phases,

$$\sigma_C = \sigma_\alpha V_\alpha + \sigma_\beta V_\beta$$

The stresses in the individual phases are given by Hooke's Law,

$$\sigma_\beta = E_\beta \epsilon$$

$$\sigma_\alpha = E_\alpha \epsilon$$

Combining these equations gives that the overall stress in the composite is,

$$\sigma_C = E_\alpha V_\alpha \epsilon + E_\beta V_\beta \epsilon = \left(E_\alpha V_\alpha + E_\beta V_\beta\right) \epsilon$$

Now we can show,

$$E_C = (E_\alpha V_\alpha + E_\beta V_\beta)$$

The lower bound is dictated by the isostress condition, in which the fibres and matrix are oriented perpendicularly to the loading direction:

$$\sigma_C = \sigma_\alpha = \sigma_\beta = \sigma$$

and now the strains become a weighted average,

$$\epsilon_C = \epsilon_\alpha V_\alpha + \epsilon_\beta V_\beta$$

Rewriting Hooke's Law for the individual phases,

$$\epsilon_\beta = \frac{\sigma}{E_\beta}$$

$$\epsilon_{\alpha} = \frac{\sigma}{E_{\alpha}}$$

This leads to,

$$\epsilon_{c} = V_{\beta}\frac{\sigma}{\epsilon_{\beta}} + V_{\alpha}\frac{\sigma}{\epsilon_{\alpha}} = (\frac{V_{\alpha}}{\epsilon_{\alpha}} + \frac{V_{\beta}}{\epsilon_{\beta}})\sigma$$

From the definition of Hooke's Law,

$$\frac{1}{E_C} = \frac{V_{\alpha}}{E_{\alpha}} + \frac{V_{\beta}}{E_{\beta}}$$

and in general,

$$\frac{1}{E_C} \sum \frac{V_1}{E_i}$$

Following the example above, if one had a composite material made up of α and β phases under isostress conditions as shown in the figure to the right, the composition Young's modulus would be:

$$E_C = (E_{\alpha}E_{\beta})/(V_{\alpha}E_{\beta} + V_{\beta}E_{\alpha})$$

The isostrain condition implies that under an applied load, both phases experience the same strain but will feel different stress. Comparatively, under isostress conditions both phases will feel the same stress but the strains will differ between each phase. Though composite stiffness is maximized when fibres are aligned with the loading direction, so is the possibility of fibre tensile fracture, assuming the tensile strength exceeds that of the matrix. When a fibre has some angle of misorientation θ, several fracture modes are possible. For small values of θ the stress required to initiate fracture is increased by a factor of $(\cos\theta)^{-2}$ due to the increased cross-sectional area ($A \cos\theta$) of the fibre and reduced force ($F/\cos\theta$) experienced by the fibre, leading to a composite tensile strength of $\sigma_{parallel}/\cos^2\theta$ where $\sigma_{parallel}$ is the tensile strength of the composite with fibres aligned parallel with the applied force.

Intermediate angles of misorientation θ lead to matrix shear failure. Again the cross sectional area is modified but since shear stress is now the driving force for failure the area of the matrix parallel to the fibres is of interest, increasing by a factor of 1/sin θ. Similarly, the force parallel to this area again decreases ($F/\cos\theta$) leading to a total tensile strength of $\tau_{my}/\sin\theta\cos\theta$ where τ_{my} is the matrix shear strength.

Finally, for large values of θ (near π/2) transverse matrix failure is the most likely to occur, since the fibres no longer carry the majority of the load. Still, the tensile strength will be greater than for the purely perpendicular orientation, since the force perpendicular to the fibres will decrease by a factor of 1/sin θ and the area decreases by a factor of 1/sin θ producing a composite tensile strength of $\sigma_{perp}/\sin^2\theta$ where σ_{perp} is the tensile strength of the composite with fibres align perpendicular to the applied force.

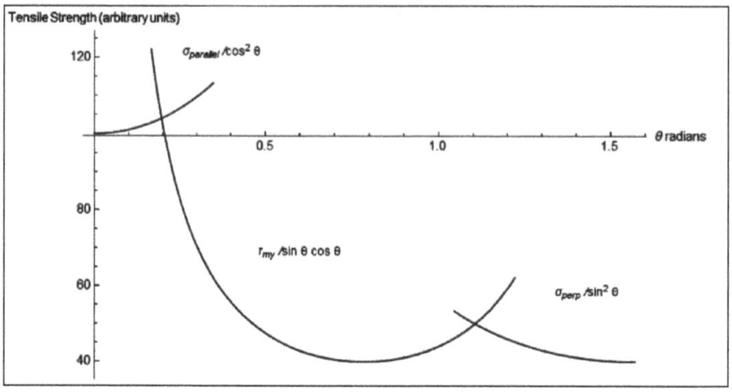

The graph depicts the three fracture modes a composite material may experience depending on the angle of misorientation relative to aligning fibres parallel to the applied stress.

The majority of commercial composites are formed with random dispersion and orientation of the strengthening fibres, in which case the composite Young's modulus will fall between the isostrain and isostress bounds. However, in applications where the strength-to-weight ratio is engineered to be as high as possible (such as in the aerospace industry), fibre alignment may be tightly controlled.

Panel stiffness is also dependent on the design of the panel. For instance, the fibre reinforcement and matrix used, the method of panel build, thermoset versus thermoplastic, and type of weave.

In contrast to composites, isotropic materials (for example, aluminium or steel), in standard wrought forms, typically have the same stiffness regardless of the directional orientation of the applied forces and/or moments. The relationship between forces/moments and strains/curvatures for an isotropic material can be described with the following material properties: Young's Modulus, the shear Modulus and the Poisson's ratio, in relatively simple mathematical relationships. For the anisotropic material, it requires the mathematics of a second order tensor and up to 21 material property constants. For the special case of orthogonal isotropy, there are three different material property constants for each of Young's Modulus, Shear Modulus and Poisson's ratio—a total of 9 constants to describe the relationship between forces/moments and strains/curvatures.

Techniques that take advantage of the anisotropic properties of the materials include mortise and tenon joints (in natural composites such as wood) and Pi Joints in synthetic composites.

Failure

Shock, impact, or repeated cyclic stresses can cause the laminate to separate at the interface between two layers, a condition known as delamination. Individual fibres can separate from the matrix e.g. fibre pull-out.

Composites can fail on the microscopic or macroscopic scale. Compression failures can occur at both the macro scale or at each individual reinforcing fibre in compression buckling. Tension failures can be net section failures of the part or degradation of the composite at a microscopic scale where one or more of the layers in the composite fail in tension of the matrix or failure of the bond between the matrix and fibres.

Some composites are brittle and have little reserve strength beyond the initial onset of failure while others may have large deformations and have reserve energy absorbing capacity past the onset of damage. The variations in fibres and matrices that are available and the mixtures that can be made with blends leave a very broad range of properties that can be designed into a composite structure. The best known failure of a brittle ceramic matrix composite occurred when the carbon-carbon composite tile on the leading edge of the wing of the Space Shuttle Columbia fractured when impacted during take-off. It led to catastrophic break-up of the vehicle when it re-entered the Earth's atmosphere on 1 February 2003.

Compared to metals, composites have relatively poor bearing strength.

Testing

To aid in predicting and preventing failures, composites are tested before and after construction. Pre-construction testing may use finite element analysis (FEA) for ply-by-ply analysis of curved surfaces and predicting wrinkling, crimping and dimpling of composites. Materials may be tested during manufacturing and after construction through several nondestructive methods including ultrasonics, thermography, shearography and X-ray radiography, and laser bond inspection for NDT of relative bond strength integrity in a localized area.

Crystal Structure of Composite Materials

A composite material is basically a combination of two or more materials, each of which retains it own distinctive properties. Multiphase metals are composite materials on a micro scale, but generally the term composite is applied to materials that are created by mechanically bonding two or more different materials together. The resulting material has characteristics that are not characteristic of the components in isolation. The concept of composite materials is ancient. An example is adding straw to mud for building stronger mud walls. Most commonly, composite materials have a bulk phase, which is continuous, called the matrix; and a dispersed, non-continuous, phase called the reinforcement. Some other examples of basic composites include concrete (cement mixed with sand and aggregate), reinforced concrete (steel rebar in concrete), and fiberglass (glass strands in a resin matrix).

In about the mid 1960's, a new group of composite materials, called advanced engineered composite materials (aka advanced composites), began to emerge. Advanced composites utilize a combination of resins and fibers, customarily carbon/graphite, kevlar, or fiberglass with an epoxy resin. The fibers provide the high stiffness, while the surrounding polymer resin matrix holds the structure together. The fundamental design concept of composites is that the bulk phase accepts the load over a large surface area, and transfers it to the reinforcement material, which can carry a greater load. The significance here lies in that there are numerous matrix materials and as many fiber types, which can be combined in countless ways to produce just the desired properties. These materials were first developed for use in the aerospace industry because for certain application they have a higher stiffness to weight or strength-to-weight ratio than metals. This means metal parts can be replaced with lighter weight parts manufactured from advanced composites. Generally, carbon-epoxy composites are two thirds the weight of aluminum, and two and a half times as stiff. Composites are resistant to fatigue damage and harsh environments, and are repairable.

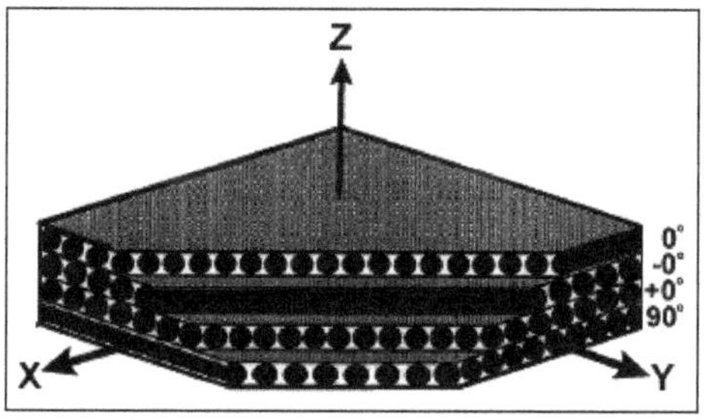

Composites meeting the criteria of having mechanical bonding can also be produced on a micro scale. For example, when tungsten carbide powder is mixed with cobalt powder, and then pressed and sintered together, the tungsten carbide retains its identity. The resulting material has a soft cobalt matrix with tough tungsten carbide particles inside. This material is used to produce carbide drill bits and is called a metal-matrix composite. A metal matrix composite is a type of metal that is reinforced with another material to improve strength, wear or some other characteristics.

Semiconductors

A semiconductor material has an electrical conductivity value falling between that of a conductor, such as metallic copper, and an insulator, such as glass. Its resistance falls as its temperature rises; metals are the opposite. Its conducting properties may be altered in useful ways by introducing impurities ("doping") into the crystal structure. Where two differently-doped regions exist in the same crystal, a semiconductor junction is created. The behavior of charge carriers which include electrons, ions and electron holes at these junctions is the basis of diodes, transistors and all modern electronics. Some examples of semiconductors are silicon, germanium, gallium arsenide, and elements near the so-called "metalloid staircase" on the periodic table. After silicon, gallium arsenide is the second most common semiconductor and is used in laser diodes, solar cells, microwave-frequency integrated circuits and others. Silicon is a critical element for fabricating most electronic circuits.

Semiconductor devices can display a range of useful properties such as passing current more easily in one direction than the other, showing variable resistance, and sensitivity to light or heat. Because the electrical properties of a semiconductor material can be modified by doping, or by the application of electrical fields or light, devices made from semiconductors can be used for amplification, switching, and energy conversion.

The conductivity of silicon is increased by adding a small amount (of the order of 1 in 10^8) of pentavalent (antimony, phosphorus, or arsenic) or trivalent (boron, gallium, indium) atoms. This process is known as doping and resulting semiconductors are known as doped or extrinsic semiconductors. Apart from doping, the conductivity of a semiconductor can equally be improved by increasing its temperature. This is contrary to the behaviour of a metal in which conductivity decreases with increase in temperature.

The modern understanding of the properties of a semiconductor relies on quantum physics to explain the movement of charge carriers in a crystal lattice. Doping greatly increases the number of charge carriers within the crystal. When a doped semiconductor contains mostly free holes it is called "p-type", and when it contains mostly free electrons it is known as "n-type". The semiconductor materials used in electronic devices are doped under precise conditions to control the concentration and regions of p- and n-type dopants. A single semiconductor crystal can have many p- and n-type regions; the p–n junctions between these regions are responsible for the useful electronic behavior.

Some of the properties of semiconductor materials were observed throughout the mid 19th and first decades of the 20th century. The first practical application of semiconductors in electronics was the 1904 development of the cat's-whisker detector, a primitive semiconductor diode used in early radio receivers. Developments in quantum physics in turn led to the development of the transistor in 1947, the integrated circuit in 1958, and the MOSFET (metal–oxide–semiconductor field-effect transistor) in 1959.

Properties

Variable Electrical Conductivity

Semiconductors in their natural state are poor conductors because a current requires the flow of electrons, and semiconductors have their valence bands filled, preventing the entire flow of new electrons. There are several developed techniques that allow semiconducting materials to behave like conducting materials, such as doping or gating. These modifications have two outcomes: n-type and p-type. These refer to the excess or shortage of electrons, respectively. An unbalanced number of electrons would cause a current to flow through the material.

Heterojunctions

Heterojunctions occur when two differently doped semiconducting materials are joined together. For example, a configuration could consist of p-doped and n-doped germanium. This results in an exchange of electrons and holes between the differently doped semiconducting materials. The n-doped germanium would have an excess of electrons, and the p-doped germanium would have an excess of holes. The transfer occurs until equilibrium is reached by a process called recombination, which causes the migrating electrons from the n-type to come in contact with the migrating holes from the p-type. A product of this process is charged ions, which result in an electric field.

Excited Electrons

A difference in electric potential on a semiconducting material would cause it to leave thermal equilibrium and create a non-equilibrium situation. This introduces electrons and holes to the system, which interact via a process called ambipolar diffusion. Whenever thermal equilibrium is disturbed in a semiconducting material, the number of holes and electrons changes. Such disruptions can occur as a result of a temperature difference or photons, which can enter the system and create electrons and holes. The process that creates and annihilates electrons and holes are called generation and recombination.

Light Emission

In certain semiconductors, excited electrons can relax by emitting light instead of producing heat. These semiconductors are used in the construction of light-emitting diodes and fluorescent quantum dots.

High Thermal Conductivity

Semiconductors with high thermal conductivity can be used for heat dissipation and improving thermal management of electronics.

Thermal Energy Conversion

Semiconductors have large thermoelectric power factors making them useful in thermoelectric generators, as well as high thermoelectric figures of merit making them useful in thermoelectric coolers.

Materials

A large number of elements and compounds have semiconducting properties, including:

- Certain pure elements are found in Group 14 of the periodic table; the most commercially important of these elements are silicon and germanium. Silicon and germanium are used here effectively because they have 4 valence electrons in their outermost shell which gives them the ability to gain or lose electrons equally at the same time.

- Binary compounds, particularly between elements in Groups 13 and 15, such as gallium arsenide, Groups 12 and 16, groups 14 and 16, and between different group 14 elements, e.g. silicon carbide.
- Certain ternary compounds, oxides and alloys.
- Organic semiconductors, made of organic compounds.

Silicon crystals are the most common semiconducting materials used in microelectronics and photovoltaics.

Most common semiconducting materials are crystalline solids, but amorphous and liquid semiconductors are also known. These include hydrogenated amorphous silicon and mixtures of arsenic, selenium and tellurium in a variety of proportions. These compounds share with better known semiconductors the properties of intermediate conductivity and a rapid variation of conductivity with temperature, as well as occasional negative resistance. Such disordered materials lack the rigid crystalline structure of conventional semiconductors such as silicon. They are generally used in thin film structures, which do not require material of higher electronic quality, being relatively insensitive to impurities and radiation damage.

Preparation of Semiconductor Materials

Almost all of today's electronic technology involves the use of semiconductors, with the most important aspect being the integrated circuit (IC), which are found in laptops, scanners, cell-phones, etc. Semiconductors for ICs are mass-produced. To create an ideal semiconducting material, chemical purity is paramount. Any small imperfection can have a drastic effect on how the semiconducting material behaves due to the scale at which the materials are used.

A high degree of crystalline perfection is also required, since faults in crystal structure (such as dislocations, twins, and stacking faults) interfere with the semiconducting properties of the material. Crystalline faults are a major cause of defective semiconductor devices. The larger the crystal, the more difficult it is to achieve the necessary perfection. Current mass production processes use crystal ingots between 100 and 300 mm (3.9 and 11.8 in) in diameter which are grown as cylinders and sliced into wafers.

There is a combination of processes that is used to prepare semiconducting materials for ICs. One process is called thermal oxidation, which forms silicon dioxide on the surface of the silicon. This is used as a gate insulator and field oxide. Other processes are called photomasks and photolithography. This process is what creates the patterns on the circuity in the integrated circuit. Ultraviolet light is used along with a photoresist layer to create a chemical change that generates the patterns for the circuit.

Etching is the next process that is required. The part of the silicon that was not covered by the photoresist layer from the previous step can now be etched. The main process typically used today is called plasma etching. Plasma etching usually involves an etch gas pumped in a low-pressure chamber to create plasma. A common etch gas is chlorofluorocarbon, or more commonly known Freon. A high radio-frequency voltage between the cathode and anode is what creates the plasma in the chamber. The silicon wafer is located on the cathode, which causes it to be hit by the positively charged ions that are released from the plasma. The end result is silicon that is etched anisotropically.

The last process is called diffusion. This is the process that gives the semiconducting material its desired semiconducting properties. It is also known as doping. The process introduces an impure atom to the system, which creates the p-n junction. In order to get the impure atoms embedded in the silicon wafer, the wafer is first put in a 1,100 degree Celsius chamber. The atoms are injected in and eventually diffuse with the silicon. After the process is completed and the silicon has reached room temperature, the doping process is done and the semiconducting material is ready to be used in an integrated circuit.

Physics of Semiconductors

Energy Bands and Electrical Conduction

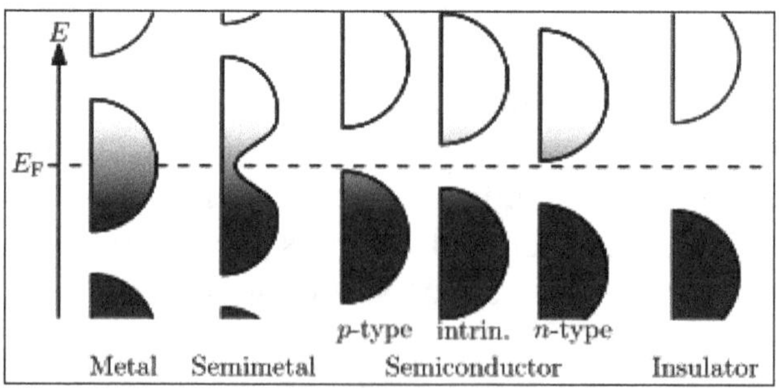

Filling of the electronic states in various types of materials at equilibrium. Here, height is energy while width is the density of available states for a certain energy in the material listed. The shade follows the Fermi–Dirac distribution (black = all states filled, white = no state filled). In metals and semimetals the Fermi level E_F lies inside at least one band. In insulators and semiconductors the Fermi level is inside a band gap; however, in semiconductors the bands are near enough to the Fermi level to be thermally populated with electrons or holes.

Semiconductors are defined by their unique electric conductive behavior, somewhere between that of a conductor and an insulator. The differences between these materials can be understood in terms of the quantum states for electrons, each of which may contain zero or one electron (by the Pauli exclusion principle). These states are associated with the electronic band structure of the material. Electrical conductivity arises due to the presence of electrons in states that are delocalized (extending through the material), however in order to transport electrons a state must be *partially filled*, containing an electron only part of the time. If the state is always occupied with an electron, then it is inert, blocking the passage of other electrons via that state. The energies of these quantum states are critical, since a state is partially filled only if its energy is near the Fermi level.

High conductivity in a material comes from it having many partially filled states and much state delocalization. Metals are good electrical conductors and have many partially filled states with energies near their Fermi level. Insulators, by contrast, have few partially filled states, their Fermi levels sit within band gaps with few energy states to occupy. Importantly, an insulator can be made to conduct by increasing its temperature: heating provides energy to promote some electrons across the band gap, inducing partially filled states in both the band of states beneath the band gap (valence band) and the band of states above the band gap (conduction band). An (intrinsic) semiconductor has a band gap that is smaller than that of an insulator and at room temperature significant numbers of electrons can be excited to cross the band gap.

A pure semiconductor, however, is not very useful, as it is neither a very good insulator nor a very good conductor. However, one important feature of semiconductors (and some insulators, known as *semi-insulators*) is that their conductivity can be increased and controlled by doping with impurities and gating with electric fields. Doping and gating move either the conduction or valence band much closer to the Fermi level, and greatly increase the number of partially filled states.

Some wider-band gap semiconductor materials are sometimes referred to as semi-insulators. When undoped, these have electrical conductivity nearer to that of electrical insulators, however they can be doped (making them as useful as semiconductors). Semi-insulators find niche applications in micro-electronics, such as substrates for HEMT. An example of a common semi-insulator is gallium arsenide. Some materials, such as titanium dioxide, can even be used as insulating materials for some applications, while being treated as wide-gap semiconductors for other applications.

Charge Carriers (Electrons and Holes)

The partial filling of the states at the bottom of the conduction band can be understood as adding electrons to that band. The electrons do not stay indefinitely (due to the natural thermal recombination) but they can move around for some time. The actual concentration of electrons is typically very dilute, and so (unlike in metals) it is possible to think of the electrons in the conduction band of a semiconductor as a sort of classical ideal gas, where the electrons fly around freely without being subject to the Pauli exclusion principle. In most semiconductors the conduction bands have a parabolic dispersion relation, and so these electrons respond to forces (electric field, magnetic field, etc.) much like they would in a vacuum, though with a different effective mass. Because the electrons behave like an ideal gas, one may also think about conduction in very simplistic terms such as the Drude model, and introduce concepts such as electron mobility.

For partial filling at the top of the valence band, it is helpful to introduce the concept of an electron hole. Although the electrons in the valence band are always moving around, a completely full valence band is inert, not conducting any current. If an electron is taken out of the valence band, then the trajectory that the electron would normally have taken is now missing its charge. For the purposes of electric current, this combination of the full valence band, minus the electron, can be converted into a picture of a completely empty band containing a positively charged particle that moves in the same way as the electron. Combined with the *negative* effective mass of the electrons at the top of the valence band, we arrive at a picture of a positively charged particle that responds to electric and magnetic fields just as a normal positively charged particle would do in vacuum, again with some positive effective mass. This particle is called a hole, and the collection of holes in the valence band can again be understood in simple classical terms (as with the electrons in the conduction band).

Carrier Generation and Recombination

When ionizing radiation strikes a semiconductor, it may excite an electron out of its energy level and consequently leave a hole. This process is known as *electron–hole pair generation*. Electron-hole pairs are constantly generated from thermal energy as well, in the absence of any external energy source.

Electron-hole pairs are also apt to recombine. Conservation of energy demands that these recombination events, in which an electron loses an amount of energy larger than the band gap, be accompanied by the emission of thermal energy (in the form of phonons) or radiation (in the form of photons).

In some states, the generation and recombination of electron–hole pairs are in equipoise. The number of electron-hole pairs in the steady state at a given temperature is determined by quantum statistical mechanics. The precise quantum mechanical mechanisms of generation and recombination are governed by conservation of energy and conservation of momentum.

As the probability that electrons and holes meet together is proportional to the product of their numbers, the product is in steady state nearly constant at a given temperature, providing that there is no significant electric field (which might "flush" carriers of both types, or move them from neighbour regions containing more of them to meet together) or externally driven pair generation. The product is a function of the temperature, as the probability of getting enough thermal energy to produce a pair increases with temperature, being approximately $\exp(-E_G/kT)$, where k is Boltzmann's constant, T is absolute temperature and E_G is band gap.

The probability of meeting is increased by carrier traps – impurities or dislocations which can trap an electron or hole and hold it until a pair is completed. Such carrier traps are sometimes purposely added to reduce the time needed to reach the steady state.

Doping

The conductivity of semiconductors may easily be modified by introducing impurities into their crystal lattice. The process of adding controlled impurities to a semiconductor is known as *doping*. The amount of impurity, or dopant, added to an *intrinsic* (pure) semiconductor varies its level of conductivity. Doped semiconductors are referred to as *extrinsic*. By adding impurity to the pure semiconductors, the electrical conductivity may be varied by factors of thousands or millions.

A 1 cm^3 specimen of a metal or semiconductor has of the order of 10^{22} atoms. In a metal, every atom donates at least one free electron for conduction, thus 1 cm^3 of metal contains on the order of 10^{22} free electrons, whereas a 1 cm^3 sample of pure germanium at 20 °C contains about 4.2×10^{22} atoms, but only 2.5×10^{13} free electrons and 2.5×10^{13} holes. The addition of 0.001% of arsenic (an impurity) donates an extra 10^{17} free electrons in the same volume and the electrical conductivity is increased by a factor of 10,000.

The materials chosen as suitable dopants depend on the atomic properties of both the dopant and the material to be doped. In general, dopants that produce the desired controlled changes are classified as either electron acceptors or donors. Semiconductors doped with *donor* impurities are called *n-type*, while those doped with *acceptor* impurities are known as *p-type*. The n and p type designations indicate which charge carrier acts as the material's majority carrier. The opposite carrier is called the minority carrier, which exists due to thermal excitation at a much lower concentration compared to the majority carrier.

For example, the pure semiconductor silicon has four valence electrons which bond each silicon atom to its neighbors. In silicon, the most common dopants are *group III* and *group V* elements. Group III elements all contain three valence electrons, causing them to function as acceptors when used to dope silicon. When an acceptor atom replaces a silicon atom in the crystal, a vacant state (an electron "hole") is created, which can move around the lattice and functions as a charge carrier. Group V elements have five valence electrons, which allows them to act as a donor; substitution of these atoms for silicon creates an extra free electron. Therefore, a silicon crystal doped with boron creates a p-type semiconductor whereas one doped with phosphorus results in an n-type material.

During manufacture, dopants can be diffused into the semiconductor body by contact with gaseous compounds of the desired element, or ion implantation can be used to accurately position the doped regions.

Amorphous Semiconductors

Some materials, when rapidly cooled to a glassy amorphous state, have semiconducting properties. These include B, Si, Ge, Se, and Te, and there are multiple theories to explain them.

Crystal Structure of Semiconductors

Early semiconductors were fabricated from the element *Germanium*, but *Silicon* is preferred in most modern applications.

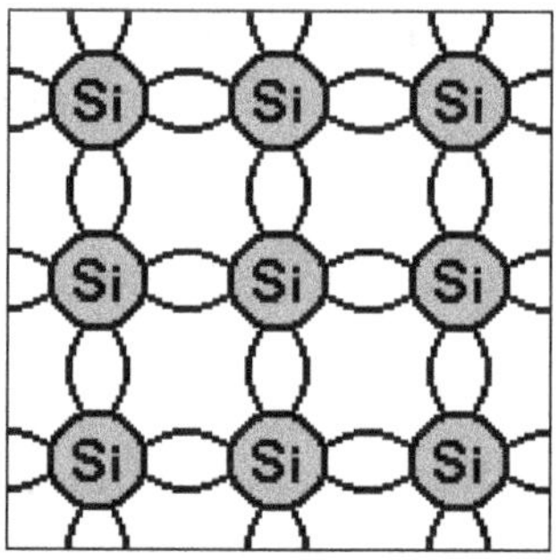

The crystal structure of pure silicon is of course 3-dimensional, but that is difficult to display or to see, so the image to the left is often used to represent the crystal structure of silicon. For you physics types, silicon (and germanium) falls in column IVa of the Periodic Table. This is the carbon family of elements. The essential characteristic of these elements is that each atom has four electrons to share with adjacent atoms in forming *bonds*.

While this is an oversimplified description, the nature of a bond between two silicon atoms is such that each atom provides one electron to share with the other. The two electrons thus shared are in fact shared equally between the two atoms. This type of sharing is known as a *covalent bond*. Such a bond is very stable, and holds the two atoms together very tightly, so that it requires a lot of energy to break this bond.

For those who are interested, the actual bonds in a 3-dimensional silicon crystal are arranged at equal angles from each other. If you visualize a tetrahedron (a pyramid with three points on the ground and a fourth point sticking straight up) with the atom centered inside, the four bonds will be directed towards the points of the tetrahedron.

Now we have our silicon crystal, but we still don't have a semiconductor. In the crystal we saw above, all of the outer electrons of all silicon atoms are used to make covalent bonds with other atoms. There are no electrons available to move from place to place as an electrical current. Thus, a pure silicon crystal is quite a good insulator. In fact, it is almost glass, which is silicon dioxide. A crystal of pure silicon is said to be an *intrinsic* crystal.

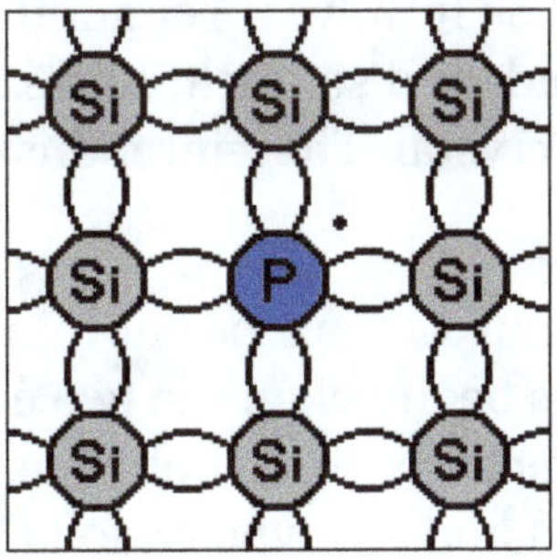

To allow our silicon crystal to conduct electricity, we must find a way to allow some electrons to move from place to place within the crystal, in spite of the covalent bonds between atoms. One way to accomplish this is to introduce an impurity such as *Arsenic* or *Phosphorus* into the crystal structure, as shown to the right. These elements are from column Va of the Periodic Table, and have five outer electrons to share with other atoms. In this application, four of these five electrons bond with adjacent silicon atoms as before, but the fifth electron cannot form a bond. This electron can easily be moved with only a small applied electrical voltage. Because the resulting crystal has an excess of current-carrying electrons, each with a negative charge, it is known as "N-type" silicon.

This construction does not conduct electricity as easily and readily as, say, copper or silver. It does exhibit some resistance to the flow of electricity. It cannot properly be called a conductor, but at the same time it is no longer an insulator. Therefore, it is known as a *semiconductor*.

While this effect is interesting, it still isn't particularly useful by itself. A plain carbon resistor is easier and cheaper to manufacture than a silicon semiconductor one. We still don't have any way to actually control an electrical current.

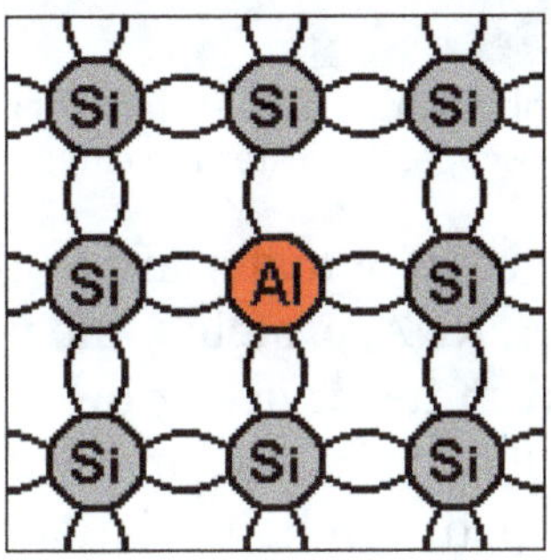

These elements only have three electrons available to share with other atoms. Those three electrons do indeed form covalent bonds with adjacent silicon atoms, but the expected fourth bond cannot be formed. A complete connection is impossible here, leaving a "hole" in the structure of the crystal.

Experimentation shows that there is an empty place where an electron should logically go, and often an electron will try to move into that space to fill it. However, the electron filling the hole had to leave a covalent bond behind to fill this empty space, and therefore leaves another hole behind as it moves. Yet another electron may move into that hole, leaving another hole behind, and so forth. In this manner, holes appear to move as positive charges through the crystal. Therefore, this type of semiconductor material is designated "P-type" silicon.

By themselves, P-type semiconductors are no more useful than N-type semiconductors. The truly interesting effects begin when the two are combined in various ways, in a single crystal of silicon. The most basic and obvious combination is a single crystal with an N-type region at one end and a P-type region at the other.

Polymers

A polymer is a chemical compound consisting of large molecules, each of which is a long chain made up of small structural units that are linked together by covalent chemical bonds. Each structural unit, called a monomer, is a small molecule of low-to-moderate molecular weight. Within a given polymer molecule, the monomers are usually identical or similar in structure. The chemical reaction by which monomers are linked together to form polymers is called polymerization.

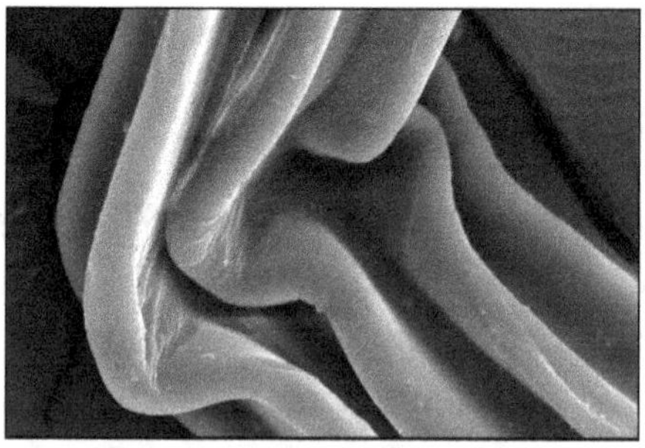

Image of a bend in a polyester fiber with a high surface area, as seen at high magnification with a scanning electron microscope. Polyester is a category of thermoplastic polymers.

Polymers form a large, diverse group of materials. Within each living organism, polymers (biopolymers) such as DNA, RNA, proteins, and polysaccharides perform specific functions that enable the organism to survive, grow, and reproduce. In addition, natural polymers—such as cotton, flax, jute, silk, and wool—have long been used for the production of clothing, rope, carpeting, felt, insulation, and upholstery. More recently, scientists have discovered how to produce new polymers with a wide range of properties, at relatively low cost. Their work has given birth to a proliferation of plastics, artificial fibers, and synthetic rubber. Consequently, synthetic polymers are being used for numerous products in homes, schools, offices, factories, recreational facilities, and means of transportation and communication. Thus, artificial polymers have become an integral part of our modern technological society.

On the downside, most artificial polymers are not biodegradable, and factories and incineration furnaces often release chemical pollutants. To help solve these problems, recycling programs have been instituted in many countries, and manufacturing plants and incinerators are now fitted with pollutant traps. In addition, biodegradable polymers are being sought out.

General Characteristics and Classification

Close-up of a polyester shirt.

Most polymers are organic—that is, their long chains have backbones of mostly carbon atoms. There are also some inorganic polymers, such as the silicones, which have a backbone of alternating silicon and oxygenatoms.

Polymer chains may or may not be cross-linked with one another. Thus the molecules of a polymer can have various topologies (shapes), such as linear (unbranched), branched, network (cross-linked 3-dimensional structure), comb, or star. The properties of a polymer depend on these shapes and on the structures of the monomers that make up the chains. For example, branched polymer chains cannot line up as close to one another as linear chains can. As a result, intermolecular bonds between branched chains are weaker, and such materials have lower densities, lower melting points, and lower tensile strength. Also, properties such as the solubility, flexibility, and strength of the polymer vary according to the types of monomers in the chains.

Polymers are typically classified as follows:

- Thermoplastics: A thermoplastic is a material that is deformable, melts to a liquid when heated, and freezes to a brittle, glassy state when cooled sufficiently. Most thermoplastics are polymers whose molecules have linear or branched structures. The molecules associate with one another through various interactions: weak van der Waals forces, as in the case of polyethylene and polypropylene;

stronger dipole-dipole interactions; hydrogen bonding, as in the case of nylon; or the stacking of aromatic rings, as in the case of polystyrene.

- Thermosets (or thermosetting plastics): These are materials that are taken through a "curing" process with the addition of energy. The energy may be in the form of heat (generally above 200 °C), a chemical reaction, or irradiation. Thermoset materials are usually liquidy, powdery, or malleable prior to curing, and designed to be molded into their final form or used as adhesives. During the curing process, molecules of the starting material become cross-linked and take on a stronger form. Once cured, the thermoset cannot be remelted and remolded. Examples of thermosets are vulcanized rubber, Bakelite™ (used in electrical insulators), melamine (used in worktop surfaces), and epoxy resin (used as an adhesive).

- Elastomers: The term elastomer is applied to an "elastic polymer"—that is, a polymer that returns to its original shape when a load is removed. Elastomers are usually thermosets (that require curing), but some are thermoplastic. The long polymer chains become cross-linked during curing and account for the flexible nature of the material. The molecular form of elastomers has been likened to a "spaghetti and meatball" structure, where the meatballs signify cross-links between the flexible spaghetti strands (polymer chains). Most elastomers are rubbers, and the term elastomer is often used interchangeably with the term rubber. Examples of thermoplastic elastomers are Hytrel and Santoprene.

- Coordination polymers: In a coordination polymer, many metal centers are interconnected through ligand bridges. Most of the common halides and oxides are coordination polymers. In a more conventional sense, the term coordination polymer is reserved for compounds where the metals are bridged by polyatomic ligands, such as cyanide and carboxylates. One of the most popular bridging ligands used in the synthesis of these polymers is a tricarboxylic acid called BTC (benzene-1,3,5-tricarboxylic acid). The polymers are metal salts of this acid. Another coordination polymer is Prussian Blue, which is based on Fe-CN-Fe linkages.

- Biopolymers (biological polymers): Biopolymers are a special class of polymers produced within living organisms. They include starch, proteins, peptides, DNA, and RNA. Their monomer units are sugars, amino acids (for proteins and peptides), and nucleotides (for DNA and RNA). Unlike synthetic (artificially produced) polymers, each biopolymer has a well-defined structure. Many biopolymers spontaneously fold into characteristic shapes that determine their biological functions.

Synthetic polymers are often named after the monomer from which they are made. For example, polyethene (also called polyethylene) is the name given to the polymer formed when thousands of ethene (ethylene) molecules are bonded together. The polyethene molecules are straight or branched chains of repeating $-CH_2-CH_2-$ units (with a $-CH_3$ at each terminus). The polymerization reaction can be written as follows.

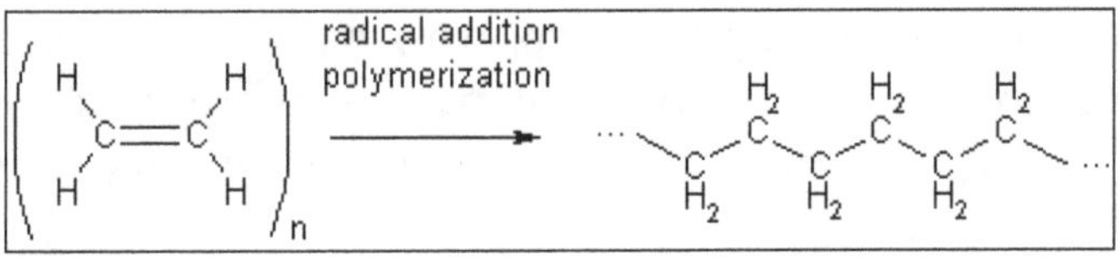

The product may also be written as:

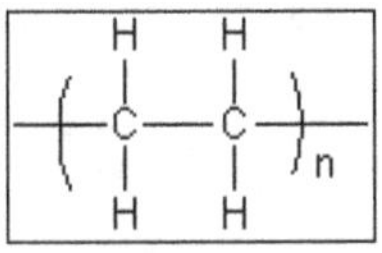

By contrast, biopolymers have been named apart from their monomeric constitution. For instance, proteins are polymers of amino acids. Typically, each protein chain is made up of hundreds of amino acid monomers, and the sequence of these monomers determines its shape and biological function.

Whereas polyethylene forms spontaneously under the right conditions, the synthesis of biopolymers such as proteins and nucleic acids requires the help of specialized biological machinery, including enzymes that catalyze the reactions. Unlike synthetic polymers, these biopolymers (other than carbohydrates) have exact sequences and lengths. Since the 1950s, catalysts have also revolutionized the development of synthetic polymers. By allowing more careful control over polymerization reactions, polymers with new properties—such as the ability to emit colored light—have been manufactured.

Copolymerization

Copolymerization involves the linking together of two or more different monomers, producing chains with varied properties. For example, a protein can be called a copolymer—one in which different amino acid monomers are linked together. Depending on the sequence of amino acids, the protein chains have different shapes and functions.

When ethene is copolymerized with small amounts of 1-hexene (or 4-methyl-1-pentene), the product is called linear low-density polyethene (LLDPE). The C_4 branches resulting from the hexene lower the density and prevent large crystalline regions from forming in the polymer, as they do in high-density polyethene (HDPE). This means that LLDPE can withstand strong tearing forces while maintaining flexibility.

The polymerization reaction may be carried out in a stepwise manner, to produce a structure with long sequences (or blocks) of one monomer alternating with long sequences of the other. The product is called a block copolymer.

In the case of some copolymers, called graft copolymers, entire chains of one kind (such as polystyrene) are made to grow out of the sides of chains of another kind (such as polybutadiene). The resultant product is less brittle and more impact-resistant. Thus, block and graft copolymers can combine the useful properties of both constituents and often behave as quasi-two-phase systems.

The formation of nylon is an example of step-growth polymerization, or condensation polymerization. The two types of monomers can have different R and R' groups, shown in the diagram below. The properties of the nylon can vary, depending on the R and R' groups in the monomers used.

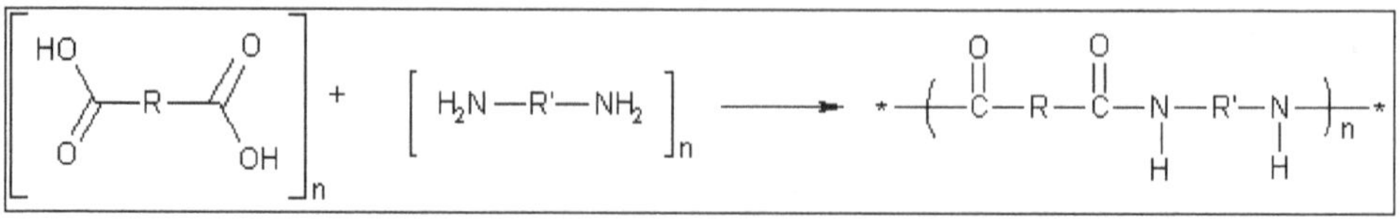

The first commercially successful, completely synthetic polymer was nylon 6,6, with four carbon atoms in the R group (adipic acid) and six carbon atoms in the R' group (hexamethylene diamine). Each monomer actually contributes 6 carbon atoms (including the two carboxyl carbons of adipic acid)—hence the name nylon 6,6. In naming nylons, the number of carbons from the diamine is given first, and the number from the diacid, second. Kevlar is an aromatic nylon in which both R and R' are benzene rings.

Copolymers illustrate the point that the repeating unit in a polymer—such as a nylon, polyester, or polyurethane—is often made up of two (or more) monomers.

Physical Properties of Polymers

Polymer chains have markedly unique physical properties, as follows:

- Molar mass distribution: During a polymerization reaction, polymer chains terminate after varying degrees of chain lengthening. The reaction produces an ensemble of differing chain lengths of differing molecular masses, with a (Gaussian) distribution around an average value. The molar mass distribution in a polymer describes this distribution of molecular masses for different chain lengths. Biopolymers, however, have well-defined structures, and they therefore do not have a molar mass distribution.
- Degree of polymerization: This is the number of monomer units in an average polymer chain, at time t in a polymerization reaction. For most industrial purposes, synthetic polymer chains need to have thousands or tens of thousands of monomer units.
- Crystallinity, and thermal phase transitions:
    - Melting point (T_m): Thermoplastic (non-cross-linked) polymers have a melting temperature above which their crystallinestructure entirely disappears.
    - Glass transition temperature (T_g): The glass transition temperature of a material is the temperature below which its molecules have little relative mobility. This temperature is usually applicable to glasses and plastics that have wholly or partially amorphous phases. Thermoplastic (non-cross-linked)

polymers have a T_g value below which they become rigid and brittle, and can crack and shatter under stress. (The T_g value is lower than T_m.) Above T_g, the polymer becomes rubbery and capable of deformation without fracture. This is one of the properties that make many plastics useful. Such behavior, however, is not exhibited by cross-linked thermosetting plastics—once cured, they are set for life, never deforming or melting when heated.

- Stereoregularity (or tacticity): This property describes the arrangement of functional groups on the backbone of carbon chains.

Chemical Properties of Polymers

The attractive forces between polymer chains play a large part in determining a polymer's properties. Given that polymer chains are so long, these interchain forces are amplified far beyond the attractions between conventional molecules. Also, longer chains are more amorphous (randomly oriented). Polymers can be visualized as tangled spaghetti chains—the more tangled the chains, the more difficult it is to pull any one strand out. These stronger forces typically result in high tensile strength and melting points.

The intermolecular forces in polymers are determined by dipoles in the monomer units. For example, polymers containing amide groups can form hydrogen bonds between adjacent chains. The somewhat positively charged hydrogen atoms in the N-H groups of one chain are strongly attracted to the somewhat negatively charged oxygen atoms in the C=O groups on another. Such strong hydrogen bonds are responsible for the high tensile strength and melting point of Kevlar.

In the case of polyesters, there is dipole-dipole bonding between the oxygen atoms in C=O groups and the hydrogen atoms in C-H groups. Dipole bonding is not as strong as hydrogen bonding, so the polyester's melting point and strength are lower than Kevlar's, but polyesters have greater flexibility.

If one considers polyethene, the monomer units (ethene) have no permanent dipole. Attractive forces between polyethene chains arise from weak van der Waals forces. Molecules can be thought of as being surrounded by a cloud of negative electrons. As two polymer chains approach, their electron clouds repel one another. This has the effect of lowering the electron density on one side of a polymer chain, creating a slight positive charge on this side. This charge is enough to attract the second polymer chain. Van der Waals forces are quite weak, however, so polyethene melts at low temperatures.

Applications of Synthetic Polymers

- Acrylonitrile butadiene styrene (ABS): This is a common thermoplastic, appropriate for making light but rigid products such as automotive body parts, protective head gear, golf club heads, and LEGO toys.

- Polyacrylates (acrylic): Noted for their transparency and resistance to breakage, polyacrylates may be used as substitutes for window glass. A familiar product in this group is Plexiglas .
- Cellulose acetate: It is used as a film base in photography, as a component in some adhesives, and as a synthetic fiber. The fiber form is used for dresses, draperies, upholstery, diapers, cigarette filters and other filters, and fiber-tip pens.
- Ionomers: These are useful for golf ball covers, semipermeable membranes, dental cements, and fuel cells.
- Liquid crystal polymers: Uses for this group of polymers include electrical and electronic applications, automotive parts, and engineering parts.
- Polyamides, such as nylon and Kevlar : Nylon fibers are used in clothing, parachutes, ropes, carpets, guitar and racket strings, and fishing nets. Kevlar is used in applications ranging from bicycles to bulletproof jackets.
- Polyesters, such as polyethylene terephthalate (PET) and polycarbonates: Polyester fibers are used to make fabrics for personal clothing, bed sheets, bedspreads, curtains, and so forth. In addition, polyesters are used to make bottles, films, liquid crystal displays, holograms, filters, and electrical insulation. Thermosetting polyester resins are commonly used as casting materials, fiberglass laminating resins, and nonmetallic auto-body fillers. Polyesters are also widely used as a finish on high-quality wooden products like guitars, pianos, and vehicle or yacht interiors.
- Polytetrafluoroethylene (Teflon): Among its many uses, it is suitable as an insulator in cables and connector assemblies and as a material for printed circuit boards (at microwave frequencies), bearings, bushings, and gears.
- Polyethylene (polyethene, PE): The polyethylenes are a widely used group of materials and are classified according to their molecular weight, density, and branching. For instance, ultra high molecular weight PE (UHMWPE) is used for can- and bottle-handling machine parts, moving parts on weaving machines, bearings, gears, artificial joints, and the newer bulletproof vests. High-density PE (HDPE) is used in making milk jugs, detergent bottles, margarine tubs, and garbage containers. Low-density PE (LDPE) is used for film wrap and plastic bags, as well as for some rigid containers.
- Melamine resin: Combined with formaldehyde, it produces a thermoset plastic that is used to make decorative wall panels, laminates, kitchen utensils, and plates. It is the main constituent of Formica and Arborite .
- Epoxy resin: It is used for many applications, including coatings, adhesives, and

composite materials, such as those using carbon fiber and fiberglass reinforcements.

- Polybutadiene (BR): This synthetic rubber has a high resistance to wear and is used mainly for the manufacture of tires.
- Polychloroprene (Neoprene): This synthetic rubber has many applications, such as for wetsuits, electrical insulation, car fan belts, gaskets, hoses, corrosion-resistant coatings, and as padding in metal cases.

Applications of Biopolymers

- Cotton: This soft fiber, which grows around the seeds of the cotton plant (Gossypium species), consists of nearly pure cellulose. It is most often spun into thread and used to make a soft, breathable textile, the most widely used natural fiber in clothing today.
- Flax: Flax fibers have been used for the production of linen for 5,000 years. The best grades are used for fabrics such as damasks, lace, and sheeting. Coarser grades are used for manufacturing twine and rope. Flax fiber is also a raw material for the high-quality paper used for banknotes.
- Hemp: Hemp fibers, obtained from the Cannabis species of plants, are used to make cordage and clothing.
- Jute: Jute fibers, composed of plant cellulose and lignin, are used to make coarse fabrics (called burlap or hessian cloth) and sacks (called gunny bags).
- Kenaf: Kenaf fibers, made by the kenaf plant (Hibiscus cannabinus), are used for the manufacture of rope, twine, coarse cloth, and paper.
- Silk: This protein fiber, obtained from the cocoons of silkworm larvae, is woven into textiles.
- Wool: This protein fiber, derived mainly from the fur of sheep and goats, is used for making clothing, carpeting, felt, insulation, and upholstery. It is also used to absorb odors and noise in heavy machinery and stereo speakers.
- Zein: This protein, found in maize, is used in the manufacture of textile fibers, biodegradable plastics, printing inks, and adhesives. It is also used as a coating for candy, nuts, fruit, and encapsulated foods and drugs.

Natural Functions of Biopolymers

- Proteins: There are different types of proteins that are involved in numerous functions in each living cell. Examples include:
    - Catalysis of biochemical reactions, carried out by numerous enzymes.

- Transport and storage of small molecules and ions.
- Immune defense, such as by forming antibodies.
- Sending and receiving signals, such as by receptorrs on cell surfaces.
- Structural support, such as components of the skin, hair, and bone.
- Coordinated motion, such as the components of muscles and molecular motors.
- Control of cell growth, such as by factors that control the synthesis of messenger RNA and proteins.

- RNA (ribonucleic acid): There are different types of RNA that carry out different functions. Examples include:
  - Messenger RNA (mRNA): Various mRNAs get their information from DNA and serve as templates for the synthesis of proteins.
  - Transfer RNA (tRNA): Specific tRNA molecules carry specific amino acids and transfer them to growing protein chains.
  - Ribosomal RNA (rRNA): rRNA molecules are part of cellular structures called ribosomes, which function as "workbenches" on which proteins are synthesized.
  - Ribozymes: These are RNA molecules that can function as enzymes, that is, they can catalyze chemical reactions.
  - Small interfering RNA (siRNA): Among their various functions, siRNAs are involved in pathways by which they interfere with the expression of specific genes.
- DNA (deoxyribonucleic acid): A constituent of the chromosomes (and organelles such as mitochondria and chloroplasts) of living cells, DNA serves as an "informational" molecule and genetic material that is inherited. Its known functions include:
  - Carrier of information for RNA structures.
  - Carrier of information for protein structures.
  - Replication, so that it can be passed down from one generation to the next.
- Polysaccharides: These large, polymeric carbohydrates occur in different types and serve various functions. Examples are as follows:
  - Cellulose: It is a common material that provides structure for plant cell walls.

  - Starch: It is a combination of two polysaccharides (amylose and amylopectin) and is made by plants to store excess glucose.
  - Glycogen ("animal starch"): This polysaccharide is the main storage form of glucose in animal and human cells.

Examples of thermoplastics:

- Acrylonitrile butadiene styrene (ABS).
- Celluloid.
- Cellulose acetate.
- Ethylene vinyl acetate (EVA).
- Ethylene vinyl alcohol (EVAL).
- Fluoroplastics (including polytetrafluoroethylene (PTFE), or Teflon).
- Ionomers.
- Kydex, an acrylic/PVC alloy.
- Liquid crystal polymer (LCP).
- Polyacetal (POM or Acetal).
- Polyacrylates (Acrylic or Acrylates).
- Polyacrylonitrile (PAN or Acrylonitrile).
- Polyamide (PA) (including nylon and Kevlar).
- Polyamide-imide (PAI).
- Polyaryletherketone (PAEK or Ketone).
- Polybutadiene (PBD).
- Polybutylene (PB).
- Polycyclohexylene dimethylene terephthalate (PCT).
- Polyhydroxyalkanoates (PHAs).
- Polyketone (PK).
- Polyester (including polycarbonate (PC), polyethylene terephthalate (PET), polybutylene terephthalate (PBT), polylactic acid (PLA)).
- Polyethylene (PE).

- Polyetheretherketone (PEEK).
- Polyetherimide (PEI).
- Polyethersulfone (PES).
- Polyethylenechlorinates (PEC).
- Polyimide (PI).
- Polymethylpentene (PMP).
- Polyphenylene oxide (PPO).
- Polyphenylene sulfide (PPS).
- Polyphthalamide (PPA).
- Polypropylene (PP).
- Polystyrene (PS).
- Polysulfone (PSU).
- Polyvinyl chloride (PVC).
- Spectralon.

Examples of Thermosets

- Vulcanized rubber.
- Bakelite (a phenol formaldehyde resin, used in electrical insulators and plastic wear).
- Duroplast.
- Urea-formaldehyde foam (used in plywood, particleboard, and medium-density fibreboard).
- Melamine resin (used on worktop surfaces).
- Polyester resin (used in glass-reinforced plastics/fiberglass).
- Epoxy resin (used as an adhesive and in fibre-reinforced plastics such as glass-reinforced plastic and graphite-reinforced plastic).

Examples of Elastomers

Unsaturated rubbers that can be cured by sulfur vulcanization:

- Natural rubber (NR).

- Polyisoprene (IR):
    - Butyl rubber (copolymer of isobutylene and isoprene, IIR).
    - Halogenated butyl rubbers: chloro butyl rubber (CIIR), bromo butyl rubber (BIIR).
- Polybutadiene (BR):
    - Styrene-butadiene rubber (SBR, copolymer of polystyrene and polybutadiene).
    - Nitrile rubber (NBR, copolymer of polybutadiene and acrylonitrile), also called buna N rubbers.
    - Hydrated nitrile rubbers (HNBR): Therban and Zetpol.
- Chloroprene rubber (CR): polychloroprene, Neoprene, Baypren.

Saturated rubbers that cannot be cured by sulfur vulcanization:

- Ethylene propylene rubber (EPM, a copolymer of polyethylene and polypropylene).
- Ethylene propylene diene rubber (EPDM, a combination of polyethylene, polypropylene, and a diene).
- Epichlorohydrin rubber (ECO).
- Polyacrylic rubber (ACM, ABR).
- Silicone rubber (SI, Q, VMQ).
- Fluorosilicone rubber (FVMQ).
- Fluoroelastomers (FKM, FPM): Viton , Tecnoflon , Fluorel , Dai-El .
- Perfluoroelastomers (FFKM).
- Tetrafluoro ethylene/propylene rubbers (FEPM).
- Chlorosulfonated polyethylene (CSM): Hypalon.
- Ethylene-vinyl acetate (EVA).

Other Types of Elastomers

- Thermoplastic Elastomers (TPE): Hytrel , Santoprene.
- Polyurethane rubber.
- Resilin, Elastin.

- Polysulfide rubber.

Crystal Structure of Polymers

Crystalline Polymers

Highly crystalline polymers are rigid, high melting, and less affected by solvent penetration. Crystallinity makes a polymers strong, but also lowers their impact resistance. As an example, samples of polyethylene prepared under high pressure (5000 atm) have high crystallinities (95 - 99%) but are extremely brittle.

Small molecules and ions form a three-dimensional lattice with an extended regular structure that makes large crystals possible. A small portion of the NaCl lattice is modeled in the diagram at the right.

We usually describe such lattices with a unit cell - the smallest repeating unit in the lattice . In the case of NaCl, the unit cell is said to be face-centered cubic.

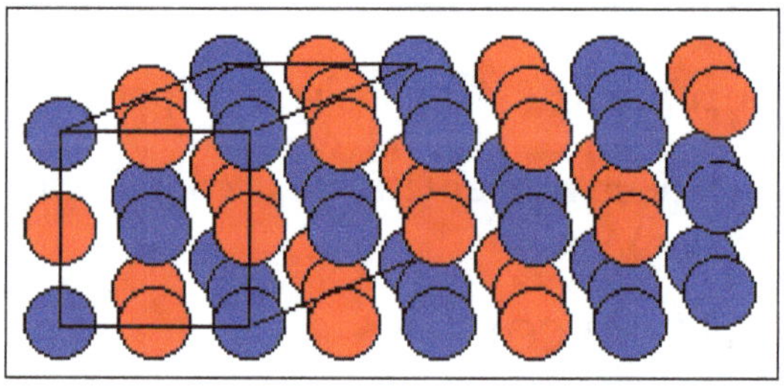

Polymer molecules are very large so it might seem that they could not pack together regularly and form a crystal. It now is known that regular polymers may form lamellar (plate-like) crystals with a thickness of 10 to 20 nm in which the parallel chains are perpendicular to the face of the crystals.

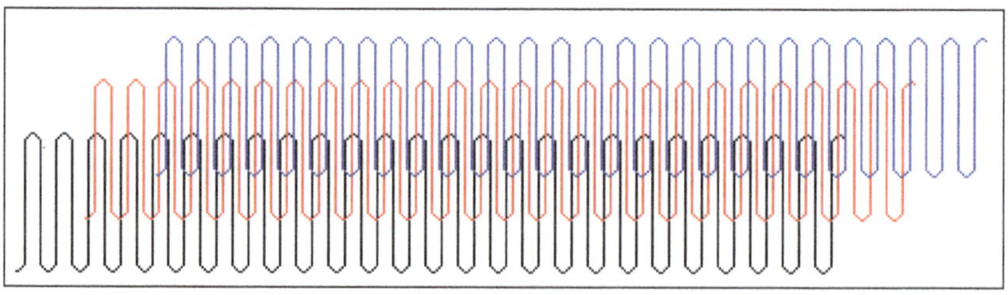

Amorphous Polymers

Polymer chains with branches or irregular pendant groups cannot pack together regularly enough to form crystals. These polymers are said to be amorphous.

Amorphous regions of a polymer are made up of a randomly coiled and entangled chains. They have been compared to a bucket containing a large number of entangled worms - each one 20-feet long and of 1/4-inch thickness[1]. The worms are so tangled that an entire worm cannot slide past the others, but small portions of the worms can twist around within the mass.

Amorphous polymers are softer, have lower melting points, and are penetrated more by solvents than are their crystalline counterparts.

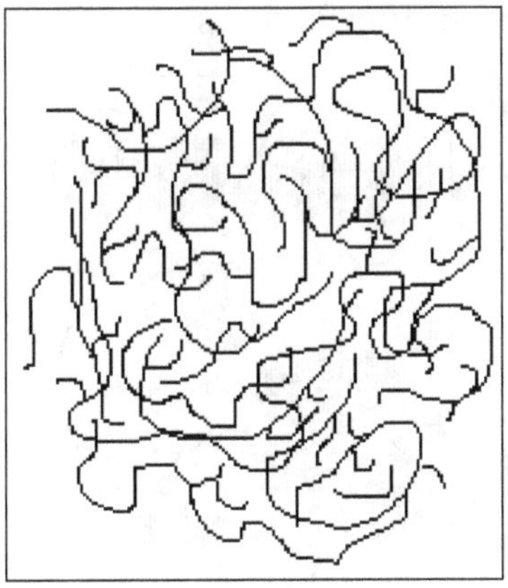

Glassy and Rubber States

At low temperatures molecular motion in an amorphous region is restricted to molecular vibrations, but the chains cannot rotate or move in space (the worms are frozen and cannot move). This form is the glassy state of the amorphous region. The glassy state can be thought of as being a supercooled liquid where the molecular motions have been frozen in. The glassy state is hard, rigid, and brittle like a crystalline solid, but retains the molecular disorder of a liquid.

When the material is heated the polymer will reach a temperature at which segments (20 - 50 atoms long) of the entangled chains can move (small portions of the entangled worms can move around so the mass in the bucket writhes). At this temperature, called the glass transition temperature, the amorphous region becomes rubbery. When an amorphous polymer is in its rubbery state it is soft and flexible.

Semi-crystalline Polymers

Semi-crystalline polymers have both crystalline and amorphous regions. Semi-crystallinity is a desirable property for most plastics because they combine the strength of crystalline polymers with the flexibility of amorphous. Semi-crystalline polymers can be tough with an ability to bend without breaking,

Isolated lamellar single crystals are obtained by crystallization from dilute solution. When crystals are formed from the melt, chain entanglements are extremely important. In this case the solid is more irregular with polymer chains meandering in and out of ordered crystalline portions. The crystalline portion is in the lamellae; the amorphous portion is outside the lamallae.

Polymers such as this are said to be semi-crystalline. The crystals are small and connected to the amorphous regions by polymer chains so there may be no sharp well-defined boundaries between the two types of regions. For some polymers, such as polyvinyl alcohol, there is a fairly distinct separation between the crystalline and amorphous regions, though. In other cases (e.g., HDPE) the structure basically is crystalline with uniformly-distributed flaws and imperfections.

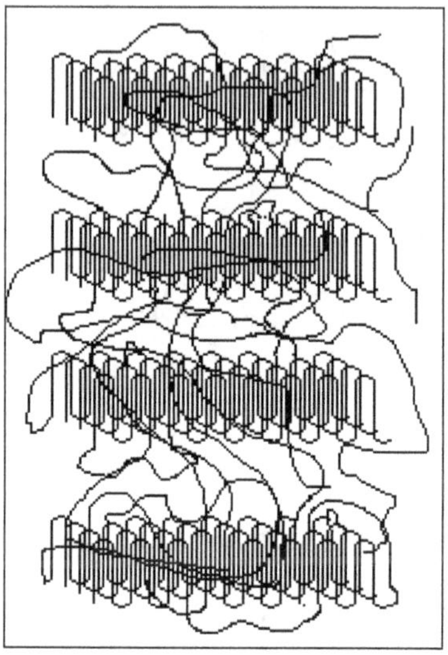

If we model a polymer as having distinct crystalline and amorphous regions then the percentage of the polymer that is crystalline is called the percent crystallinity. The percent crystallinity has an important influence on the properties of the polymer.

Spherulites

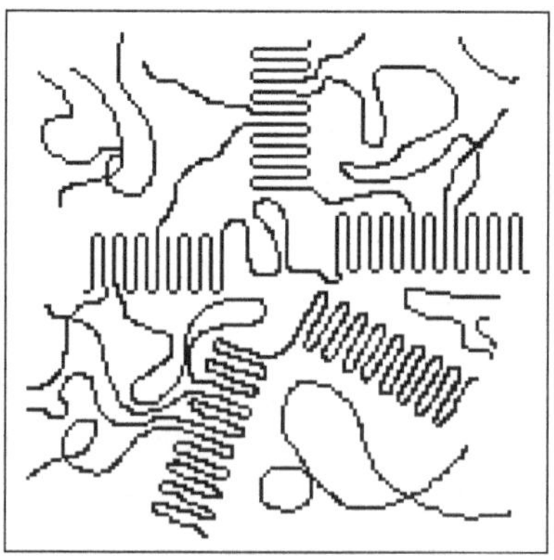

When a molten crystallizable polymer cools, the crystals grow from individual nuclei25 and radiate out like the spokes of a bicycle wheel as is illustrated in the schematic diagram at the right. The crystalline portions actually radiate out in in three dimensions, forming spheres that are called spherulites. In a sample of a crystalline polymer there are billions of spherulites.

References

- Classification-of-solids-based-on-crystal-structure, chemistry: byjus.com, Retrieved 15 May, 2020
- Wenk, Hans-Rudolf; Bulakh, Andrei (2004). Minerals: Their Constitution and Origin. Cambridge University Press. p. 10. ISBN 978-0-521-52958-7
- Metal: newworldencyclopedia.org, Retrieved 14 March, 2020
- B.G. Yacobi, Semiconductor Materials: An Introduction to Basic Principles, Springer 2003 ISBN 0-306-47361-5
- Polymer: newworldencyclopedia.org, Retrieved 12 January, 2020
- Crystalline: faculty.uscupstate.edu, Retrieved 13 April, 2020
- Carter, C. B.; Norton, M. G. (2007). Ceramic materials: Science and engineering. Springer. pp. 3 & 4. ISBN 978-0-387-46271-4
- Elhajjar, Rani; La Saponara, Valeria; Muliana, Anastasia, eds. (2017). Smart Composites: Mechanics and Design (Composite Materials). CRC Press. ISBN 978-1-138-07551-1

3
Bonding in Solids

Solids can also be classified as per the nature of the bonding between their molecular and atomic components. Four major types of bonding are covalent bonding, ionic bonding, metallic bonding and inter-molecular bonding. This chapter closely examines different types of bonding in solids to provide an extensive understanding of the subject.

Covalent Bonding

Covalent bonds are the most common and most important kind of bonding. It is a bonding between atoms within a molecule and forms the strongest bonds anywhere. Covalent bonds are chemical bonds between two non-metal atoms. A covalent bond between atoms is formed, when they share one or more pairs of electrons among each other. These shared pairs create a bond between the atoms, which binds them together into a singular unit, as a molecule. These shared electrons revolve in their outermost shells. A full outer shell needs eight electrons. The electrons in this outer shell are called valence electrons. The number of valence electrons is dependent on the size of the atom. Electrons then circle the atom in the same manner that planets orbit the stars. There is developed a path around the atomic nucleus. The first layer always contains only two electrons. The outer layers after that layer usually contain up to eight. For example, if an atom contained eight electrons, the first two would orbit very close to the atomic nucleus. The next six would orbit a little further away. Every atom wants a full outer shell. It desires to have eight electrons in the outer most shell. In this example the atom wants eight electrons in the outer shell but only has six. Therefore, the atom will borrow electrons from another atom. When this chemical bond takes place, a covalent bond is formed.

Covalent bonds can be split into two groups: Simple and Giant. Although they are both similar, there are some distinguishing factors. For example, Giant covalent bonds have relatively high melting and boiling points, whereas Simple covalent bonds have a relatively low melting and boiling point. This is because of the inter-molecular forces of attraction. With giant covalent bonds, the bonds between layers are very strong. However, with a simple covalent bond, the bonds between the atoms have a strong inter-molecular bond, but there are no bonds between the molecules, which is why it has a relatively low melting and boiling point.

When two atoms that are sharing electrons attract electrons with the same force, a non-polar covalent bond is formed. A non-polar covalent bond is a bond where the shared electrons are shared equally. On the other hand, a polar covalent bond is a covalent bond in which the shared electrons are more attracted to one of the atoms than the other. The shred electrons are more likely to be near the atom whose electronegativity is higher.

Examples of Covalent Bond

Water

An example is water. Water consists of a covalent bond containing hydrogen and oxygen bonding together to make H_2O. In this atomic molecule, two hydrogen atoms share their single electrons with the oxygen atom, which shares its own two electrons in return. This is an example of a covalent bond, which is created because of the higher electronegativity of oxygen.

Diamonds

A diamond is an example of Giant Covalent bond of carbon. A diamond has a giant molecular structure. Each carbon atom is covalently bonded to four other carbon atoms. Electrons are borrowed from these other carbon atoms. There is a tremendous amount of energy needed to separate the atoms in a diamond. This is because a covalent bond is strong and a diamond contains four covalent bonds. This makes the melting and boiling point if the diamond very high.

Vulcanized Rubber

Another example is vulcanized rubber. Rubber is sticky when warm and brittle when cold. In 1939, Charles Goodyear accidently dropped a mixture of sulfur and natural rubber on a hot stove. The mixture heated up and became tough and elastic, forming vulcanized rubber. What happened? The covalent bonds transformed sulfur and natural rubber into the vulcanized rubber when it was heated. It changed because the covalent bonds between sulfur and rubber changed.

Octet Rule

The Octet Rule requires all atoms in a molecule to have 8 valence electrons--either by sharing, losing or gaining electrons--to become stable. For Covalent bonds, atoms tend to share their electrons with each other to satisfy the Octet Rule. It requires 8 electrons because that is the amount of electrons needed to fill a *s*- and *p*- orbital (electron configuration); also known as a noble gas configuration. Each atom wants to become as stable as the noble gases that have their outer valence shell filled because noble gases have a charge of 0. Although it is important to remember the "magic number", 8, note that there are many Octet rule exceptions.

Example: As you can see from the picture below, Phosphorus has only 5 electrons in its outer shell (bolded in red). Argon has a total of 8 electrons (bolded in red), which satisfies the Octet Rule. Phosphorus needs to gain 3 electrons to fullfill the Octet Rule. It wants to be like Argon who has a full outer valence shell.

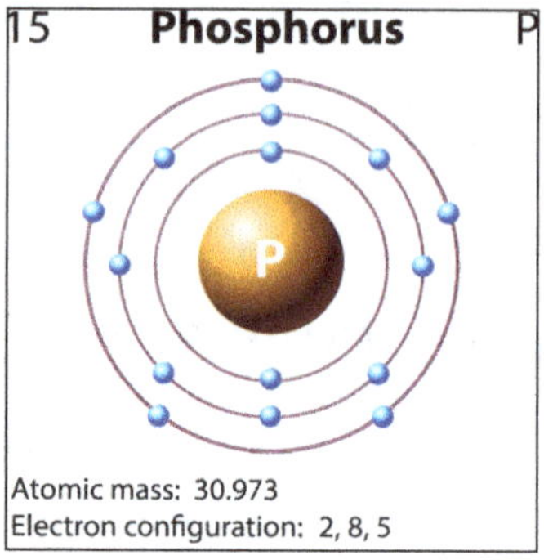

Single Bonds

A single bond is when two electrons--one pair of electrons--are shared between two atoms. It is depicted by a single line between the two atoms. Although this form of bond is weaker and has a smaller density than a double bond and a triple bond, it is the most stable because it has a lower level of reactivity meaning less vulnerability in losing electrons to atoms that want to steal electrons.

Example: HCL

Below is a Lewis dot structure of Hydrogen Chloride demonstrating a single bond. As we can see from the picture below, Hydrogen Chloride has 1 Hydrogen atom and 1 Chlorine atom. Hydrogen has only 1 valence electron whereas Chlorine has 7 valence electrons. To satisfy the Octet Rule, each atom gives out 1 electron to share with each other; thus making a single bond.

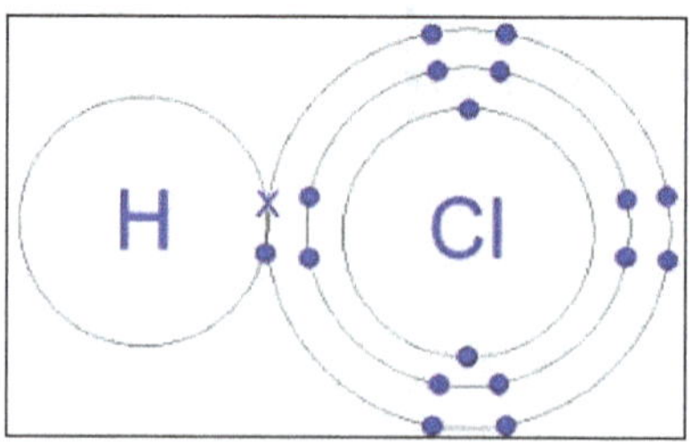

Double Bonds

A Double bond is when two atoms share two pairs of electrons with each other. It is depicted by two horizontal lines between two atoms in a molecule. This type of bond is much stronger than a single bond, but less stable; this is due to its greater amount of reactivity compared to a single bond.

Example: CO_2

Below is a Lewis dot structure of Carbon dioxide demonstrating a double bond. As you can see from the picture below, Carbon dioxide has a total of 1 Carbon atom and 2 Oxygen atoms. Each Oxygen atom has 6 valence electrons whereas the Carbon atom only has 4 valence electrons. To satisfy the Octet Rule, Carbon needs 4 more valence electrons. Since each Oxygen atom has 3 lone pairs of electrons, they can each share 1 pair of electrons with Carbon; as a result, filling Carbon's outer valence shell (Satisfying the Octet Rule).

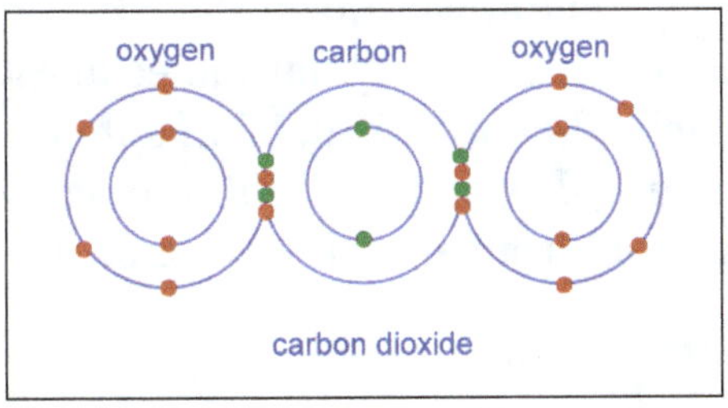

Triple Bond

A Triple bond is when three pairs of electrons are shared between two atoms in a molecule. It is the least stable out of the three general types of covalent bonds. It is very vulnerable to electron thieves.

Example: Acetylene

Below is a Lewis dot structure of Acetylene demonstrating a triple bond. As you can see from the picture below, Acetylene has a total of 2 Carbon atoms and 2 Hydrogen atoms. Each Hydrogen atom has 1 valence electron whereas each Carbon atom has 4 valence electrons. Each Carbon needs 4 more electrons and each Hydrogen needs 1 more electron. Hydrogen shares its only electron with Carbon to get a full valence shell. Now Carbon has 5 electrons. Because each Carbon atom has 5 electrons--1 single bond and 3 unpaired electrons--the two Carbons can share their unpaired electrons, forming a triple bond. Now all the atoms are happy with their full outer valence shell.

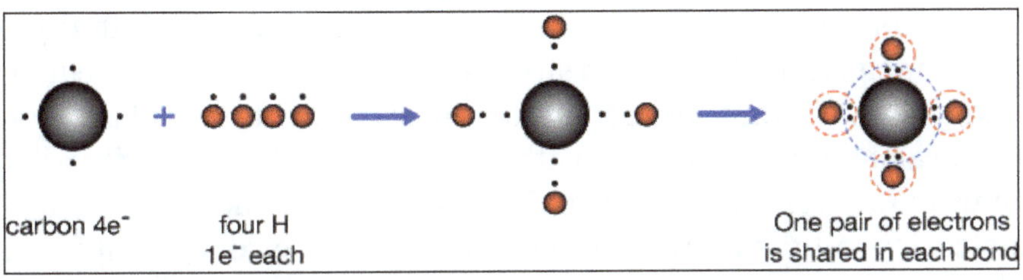

Polar Covalent Bond

A Polar Covalent Bond is created when the shared electrons between atoms are not equally shared. This occurs when one atom has a higher electronegativity than the atom it is sharing with. The atom with the higher electronegativity will have a stronger pull for electrons (Similiar to a Tug-O-War game, whoever is stronger usually wins). As a result, the shared electrons will be closer to the atom with the higher electronegativity, making it unequally shared. A polar covalent bond will result in the molecule having a slightly positive side (the side containing the atom with a lower electronegativity) and a slightly negative side (containing the atom with the higher electronegativity) because the shared electrons will be displaced toward the atom with the higher electronegativity. As a result of polar covalent bonds, the covalent compound that forms will have an electrostatic potential. This potential will make the resulting molecule slightly polar, allowing it to form weak bonds with other polar molecules. One example of molecules forming weak bonds with each other as a result of an unbalanced electrostatic potential is hydrogen bonding, where a hydrogen atom will interact with an electronegative hydrogen, fluorine, or oxygen atom from another molecule or chemical group.

Example: Water, Sulfide, Ozone, etc.

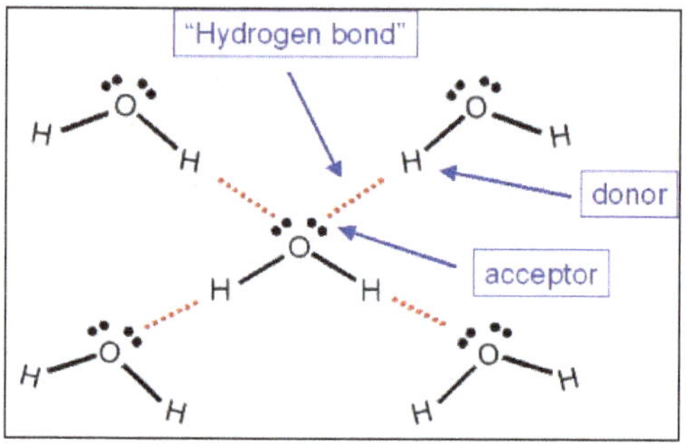

As you can see from the picture above, Oxygen is the big buff creature with the tattoo of "O" on its arm. The little bunny represents a Hydrogen atom. The blue and red bow tied in the middle of the rope, pulled by the two creatures represents--the shared pair of electrons--a single bond. Because the Hydrogen atom is weaker, the shared pair of electrons will be pulled closer to the Oxygen atom.

Nonpolar Covalent Bond

A Nonpolar Covalent Bond is created when atoms share their electrons equally. This usually occurs when two atoms have similar or the same electron affinity. The closer the values of their electron affinity, the stronger the attraction. This occurs in gas molecules; also known as diatomic elements. Nonpolar covalent bonds have a similar concept as polar covalent bonds; the atom with the higher electronegativity will draw away the electron from the weaker one. Since this statement is true--if we apply this to our diatomic molecules--all the atoms will have the same electronegativity since they are the same kind of element; thus, the electronegativities will cancel each other out and will have a charge of 0(i.e., Nonpolar covalent bond).

Examples of gas molecules that have a nonpolar covalent bond: Hydrogen gas atom, Nitrogen gas atoms, etc.

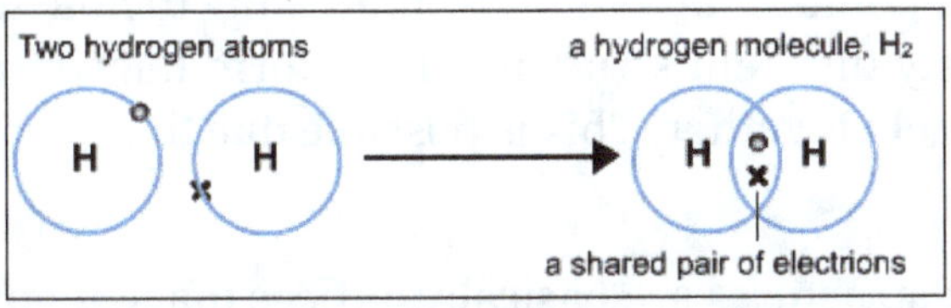

As you can see from the picture above, Hydrogen gas has a total of 2 Hydrogen atoms. Each Hydrogen atom has 1 valence electron. Since Hydrogen can only fit a max of 2 valence electrons in its orbital, each Hydrogen atom only needs 1 electron. Each atom has 1 valence electron, so they can just share, giving each atom two electrons each.

Polarization of Covalent Bonds

It is observed that in the sigma bonds between two different atoms, the electron cloud is always closer to the more electronegative of the two atoms participating in the sigma bond. Due to this, there is a permanent dipole that arises in the bond and the covalent bond is said to be polarized.

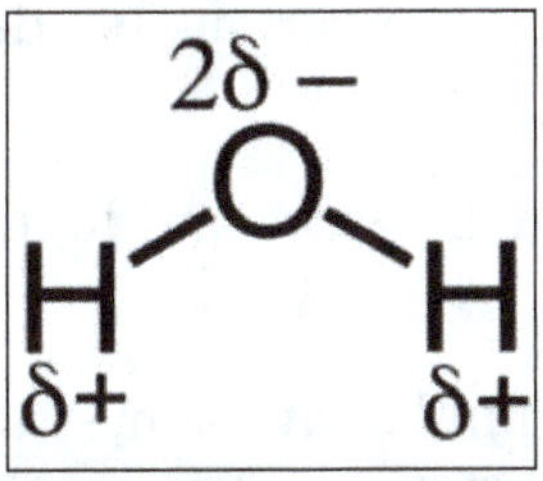

Polarity of Covalent Bond in Water Molecule.

An illustration describing the polarity of the covalent bonds in a water molecule is provided above. The more electronegative atom is said to have a partial negative charge and the less electronegative atom has a partial positive charge in the polar covalent bond.

Covalent Compounds

Covalent compounds are the ones having strong intra-molecular bonds. This is because the atoms within the covalent molecules are very tightly held together. Each molecule is indeed quite separate and the force of attraction between the individual molecules in a covalent compound tends to be weak.

We require very little energy in separating the molecules. This is because of the attractive forces between the molecules with the absence of overall electric charge. Covalent compounds are usually gaseous molecules at room temperature and pressure. They might also be liquids with low relatively low boiling points.

These characteristics could be attributed to their weak intermolecular forces which hold these atoms together. However, we also have a lot of solid covalent compounds. They have low melting points. However, it is interesting to note that a small number of these have a completely different structure. They form huge structures where a huge number of atoms are held together. This is possible due to the presence of shared electrons.

These giant molecular structures are basically lattices made up of molecules which are held together by covalent bonds structure. These covalent bonds are very strong. They also tend to be very hard with high melting points which are different from most of the covalent compounds. The example of this kind of covalent compounds includes diamond and graphite of carbon atom network. They also include silica of silicon and oxygen atoms network.

General Properties of Covalent Compounds

- Covalent compounds usually have low melting points. An exception to this include molecules of silica and diamonds that have a high melting point.
- These compounds have low boiling points. This can be attributed to their weak force of attraction between the various bonded atoms. Van Der Waals forces bind these atoms.
- These compounds are usually gases and liquids with low boiling and melting points.
- The solid covalent compounds have soft structures like graphite. This is because of the presence of a cloud of electrons in between each layer of carbons atoms.
- These compounds are non-conductors of electrical charge. The absence of charged ions is the main reason behind this. An exception to this is graphite, where we see a cloud of electrons. These make graphite a good conductor.
- They are bad conductors of heat also. Their molecules lack free electrons and that obstructs the flow of heat energy.
- Covalent compounds do not possess polar characteristics as a general property. Therefore, these compounds are insoluble in water. Water molecules are not absolutely neutral and have a slight negative charge on the oxygen atom and slight positive charges on the hydrogen atoms and since covalent compounds are made up of neutral molecules or molecules with slight charges and hence are not attracted to water molecules strongly.

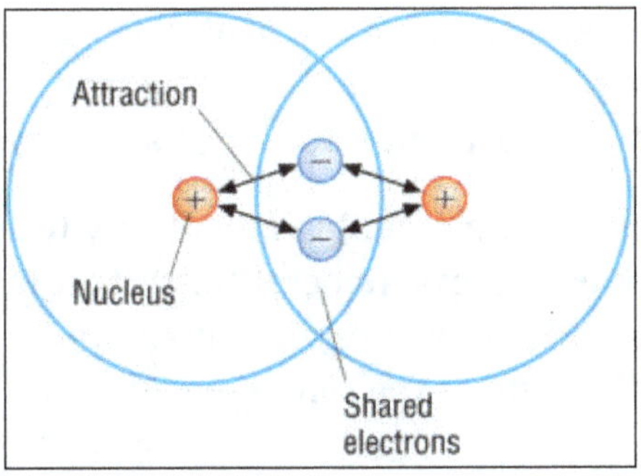

Physical and Chemical Properties

- The liquid covalent compounds evaporate. This means the molecules of liquids and solids loses from their surface into the air.
- These compounds have very less affinity between their molecules.
- Various covalent compounds have their own characteristically shaped molecules. Their bonds are directed at pre-set angles.
- Some compounds especially medicines are soluble in water. The rest are soluble in oil.
- Most of the covalent compounds are non-polar or have very little tendency to split completely to form ions and hence never conduct electricity.
- At normal temperature and pressure, we will find these compounds as either liquids or gases. But, there are solids as well and they have higher molecular weights.
- The covalent compounds crystals are of two types: One that has weak van der Waal force holding these together like in Iodine. These are easily fusible and volatile The other having a large network of atoms setting up the macromolecules.
- These compounds are soluble in organic solvents like ether and benzene.
- Covalent bonds are directional in nature. Therefore, they exhibit the phenomenon of isomerism.
- Covalent compounds majorly have a very slow rate of reactions, unlike the various ionic compounds.

Covalent Bonding of Carbon Compounds

Covalent carbon compounds are those where there is a carbon-carbon bond. These covalent compounds have stronger bonds than other compounds. This is because carbon is a small atom. Its nucleus has a strong force of attraction and holds these bonds tightly together. So covalent carbon compounds have a strong bond between themselves. Now let us understand why the covalent bonding is so relevant for carbon atoms.

As you are aware the reactivity of elements is its ability to lose or gain electrons from so that its outermost shell has a complete octet (or duplet in case of hydrogen) namely attain noble gas configuration. However, carbon has a unique situation. It has four electrons in its outermost shell, so one of the following situations must happen.

Situations:

- It can loose the four electrons in its last shell and become a cation i.e. C^{4+}. However to lose all four of these electrons would require a large amount of energy, and the resulting atom would be unstable with six neutrons holding only two electrons in one shell.
- The other option a carbon atom has is to obviously gain four electrons from another atom. But it would be extremely difficult for the resulting carbon atom to be stable. Ten electrons will have to be held by six neutrons in the nucleus.

So instead, carbon comes up with a unique solution. It shares its valence electrons with those of other carbon atoms, or even atoms of other elements. Now these shared atoms of the last shell, belong to both the atoms, hence forming a bond between these atoms. Now both atoms have a complete outer shell with eight atoms and have both attained noble gas configuration. This sharing of atoms, instead of gaining or losing is called covalent bonding. And Carbon, since its atomic number is 6, and it has four electrons in the last shell has the most favourable structure for covalent bonding.

Some Important Carbon Compounds with Covalent Bonds

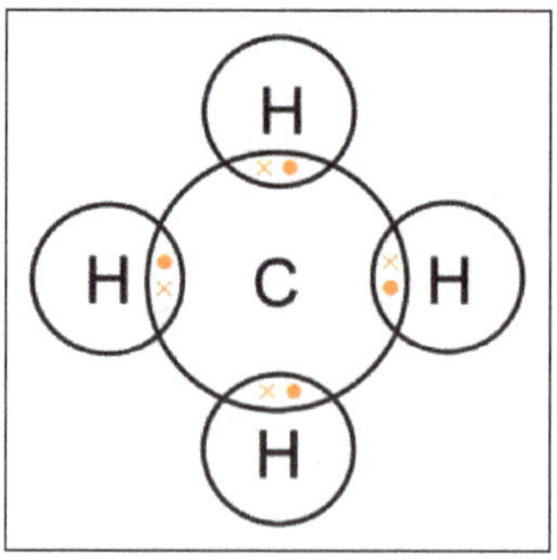

Methane: Now let us take a look at covalent bonding in some carbon elements. Let's get started with that of Methane. Its chemical formula is CH_4. Now that means one atom of carbon combines with four atoms of Hydrogen to make one molecule of methane. All the hydrogen atoms have only one electron in their outermost shell and carbon has four. So carbon shares each of its four electrons with one atom of carbon. This way carbon now has a complete octet and all four hydrogen atoms have a complete duplet.

Carbon dioxide: Let us take a look at another element you should be very familiar with, carbon dioxide. Here two atoms of oxygen combine with one atom of carbon thus giving us the molecule CO_2. As we already know now that carbon has four electrons in its outer shell, whereas oxygen with atomic number 8, has six atoms in the last shell. So carbon shares two of its atoms with each atom of oxygen. This way all three atoms complete their octet giving us one stable molecule.

Ionic Bonding

Ionic bonding is a type of chemical bonding that involves the electrostatic attraction between oppositely charged ions, and is the primary interaction occurring in ionic compounds. It is one of the main bonds along with Covalent bond and Metallic bonding. Ions are atoms that have either gained or lost one or more electrons. Ions that have gained an electron are negatively charged. They are anions. Ions that have lost an electron are positively charged. They are cations. This transfer of electrons is known as electrovalence in contrast to covalence. In the simplest case, the cation is a metal atom and the anion is a nonmetal atom, but these ions can be of a more complex nature, e.g. molecular ions like NH_4^- or SO_4^{2-}.

In simpler words, an ionic bond is the transfer of electrons from a metal to a non-metal in order to obtain a full valence shell for both atoms.

It is important to recognize that *clean* ionic bonding—in which one atom or molecule completely transfers an electron to another—cannot exist: all ionic compounds have some degree of covalent bonding, or electron sharing. Thus, the term "ionic bonding" is given when the ionic character is greater than the covalent character – that is, a bond in which a large electronegativity difference exists between the two atoms, causing the bonding to be more polar (ionic) than in covalent bonding where electrons are shared more equally. Bonds with partially ionic and partially covalent character are called polar covalent bonds.

Ionic compounds conduct electricity when molten or in solution, typically as a solid. Ionic compounds generally have a high melting point, depending on the charge of the ions they consist of. The higher the charges the stronger the cohesive forces and the higher the melting point. They also tend to be soluble in water; the stronger the cohesive forces, the lower the solubility.

Atoms that have an almost full or almost empty valence shell tend to be very reactive. Atoms that are strongly electronegative (as is the case with halogens) often have only one or two empty orbitals in their valence shell, and frequently bond with other molecules or gain electrons to form anions. Atoms that are weakly electronegative (such as alkali metals) have relatively few valence electrons, which can easily be shared with atoms that are strongly electronegative. As a result, weakly electronegative atoms tend to distort their electron cloud and form cations.

Formation

Ionic bonding can result from a redox reaction when atoms of an element (usually metal), whose ionization energy is low, give some of their electrons to achieve a stable electron configuration. In doing so, cations are formed. An atom of another element (usually nonmetal) with greater electron affinity accepts the electron(s) to attain a stable electron configuration, and after accepting electron(s) an atom becomes an anion. Typically, the stable electron configuration is one of the noble gases for elements in the s-block and the p-block, and particular stable electron configurations for d-block and f-block elements. The electrostatic attraction between the anions and cations leads to the formation of a solid with a crystallographic lattice in which the ions are stacked in an alternating fashion. In such a lattice, it is usually not possible to distinguish discrete molecular units, so that the compounds formed are not molecular in nature. However, the ions themselves can be complex and form molecular ions like the acetate anion or the ammonium cation.

For example, common table salt is sodium chloride. When sodium (Na) and chlorine (Cl) are combined, the sodium atoms each lose an electron, forming cations (Na^+), and the chlorine atoms each gain an electron to form anions (Cl^-). These ions are then attracted to each other in a 1:1 ratio to form sodium chloride (NaCl).

$$Na + Cl \rightarrow Na^+ + Cl^- \rightarrow NaCl$$

However, to maintain charge neutrality, strict ratios between anions and cations are observed so that ionic compounds, in general, obey the rules of stoichiometry despite not being molecular compounds. For compounds that are transitional to the alloys and possess mixed ionic and metallic bonding, this may not be the case anymore. Many sulfides, e.g., do form non-stoichiometric compounds.

Many ionic compounds are referred to as salts as they can also be formed by the neutralization reaction of an Arrhenius base like NaOH with an Arrhenius acid like HCl.

$$NaOH + HCl \rightarrow NaCl + H_2O$$

The salt NaCl is then said to consist of the acid rest Cl^- and the base rest Na^+.

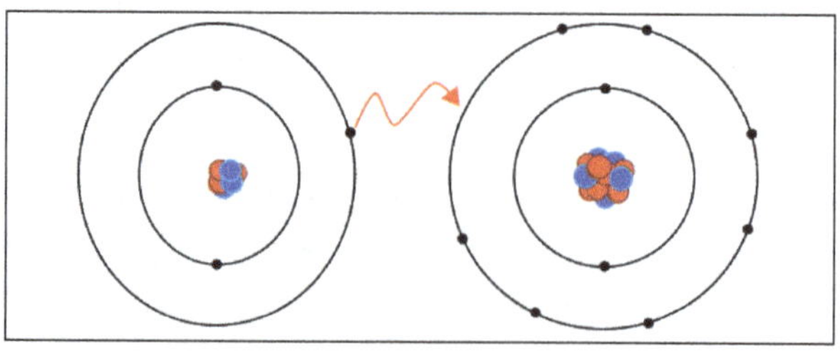

Representation of ionic bonding between lithium and fluorine to form lithium fluoride.

Lithium has a low ionization energy and readily gives up its lone valence electron to a fluorine atom, which has a positive electron affinity and accepts the electron that was donated by the lithium atom. The end-result is that lithium is isoelectronic with helium and fluorine is isoelectronic with neon. Electrostatic interaction occurs between the two resulting ions, but typically aggregation is not limited to two of them. Instead, aggregation into a whole lattice held together by ionic bonding is the result.

The removal of electrons from the cation is endothermic, raising the system's overall energy. There may also be energy changes associated with breaking of existing bonds or the addition of more than one electron to form anions. However, the action of the anion's accepting the cation's valence electrons and the subsequent attraction of the ions to each other releases (lattice) energy and, thus, lowers the overall energy of the system.

Ionic bonding will occur only if the overall energy change for the reaction is favorable. In general, the reaction is exothermic, but, e.g., the formation of mercuric oxide (HgO) is endothermic. The charge of the resulting ions is a major factor in the strength of ionic bonding, e.g. a salt $C^{+}A^{-}$ is held together by electrostatic forces roughly four times weaker than $C^{2+}A^{2-}$ according to Coulombs law, where C and A represent a generic cation and anion respectively. The sizes of the ions and the particular packing of the lattice are ignored in this rather simplistic argument.

Structures

Ionic compounds in the solid state form lattice structures. The two principal factors in determining the form of the lattice are the relative charges of the ions and their relative sizes. Some structures are adopted by a number of compounds; for example, the structure of the rock salt sodium chloride is also adopted by many alkali halides, and binary oxides such as magnesium oxide. Pauling's rules provide guidelines for predicting and rationalizing the crystal structures of ionic crystals

Strength of the Bonding

For a solid crystalline ionic compound the enthalpy change in forming the solid from gaseous ions is termed the lattice energy. The experimental value for the lattice energy can be determined using the Born–Haber cycle. It can also be calculated (predicted) using the Born–Landé equation as the sum of the electrostatic potential energy, calculated by summing interactions between cations and anions, and a short-range repulsive potential energy term. The electrostatic potential can be expressed in terms of the interionic separation and a constant (Madelung constant) that takes account of the geometry of the crystal. The further away from the nucleus the weaker the shield. The Born-Landé equation gives a reasonable fit to the lattice energy of, e.g., sodium chloride, where the calculated (predicted) value is −756 kJ/mol, which compares to −787 kJ/mol using the Born–Haber cycle. In aqueous solution the binding strength can be described by the Bjerrum or Fuoss equation as function of the ion charges, rather independent of the nature of the ions such as polaribility or size. The strength of salt bridges is most often evaluated by measurements of equilibria between molecules containing cationic and anionioc sites, most often in solution. Equilibrium constants in water indicate additive free energy contributions for each salt bridge. Another method for the identification of hydrogen bonds also in complicated molecules is crystallography, sometimes also NMR-spectroscopy.

Polarization Effects

Ions in crystal lattices of purely ionic compounds are spherical; however, if the positive ion is small and/or highly charged, it will distort the electron cloud of the negative ion, an effect summarised in Fajans' rules. This polarization of the negative ion leads to a build-up of extra charge density between the two nuclei, that is, to partial covalency. Larger negative ions are more easily polarized, but the effect is usually important only when positive ions with charges of 3+ (e.g., Al^{3+}) are involved. However, 2+ ions (Be^{2+}) or even 1+ (Li^{+}) show some polarizing power because their sizes are so small (e.g., LiI is ionic but has some covalent bonding present). This is not the ionic polarization effect that refers to displacement of ions in the lattice due to the application of an electric field.

Comparison with Covalent Bonding

In ionic bonding, the atoms are bound by attraction of oppositely charged ions, whereas, in covalent bonding, atoms are bound by sharing electrons to attain stable electron configurations. In covalent bonding, the molecular geometry around each atom is determined by valence shell electron pair repulsion VSEPR rules, whereas, in ionic materials, the geometry follows maximum packing rules. One could say that covalent bonding is more *directional* in the sense that the energy penalty for not adhering to the optimum bond angles is large, whereas ionic bonding has no such penalty. There are no shared electron pairs to repel each other, the ions should simply be packed as efficiently as possible. This often leads to much higher coordination numbers. In NaCl, each ion has 6 bonds and all bond angles are 90°. In CsCl the coordination number is 8. By comparison carbon typically has a maximum of four bonds.

Purely ionic bonding cannot exist, as the proximity of the entities involved in the bonding allows some degree of sharing electron density between them. Therefore, all ionic bonding has some covalent character. Thus, bonding is considered ionic where the ionic character is greater than the covalent character. The larger the difference in electronegativity between the two types of atoms involved in the bonding, the more ionic (polar) it is. Bonds with partially ionic and partially covalent character are called polar covalent bonds. For example, Na–Cl and Mg–O interactions have a few percent covalency, while Si–O bonds are usually ~50% ionic and ~50% covalent. Pauling estimated that an electronegativity difference of 1.7 (on the Pauling scale) corresponds to 50%

ionic character, so that a difference greater than 1.7 corresponds to a bond which is predominantly ionic. Ionic character in covalent bonds can be directly measured for atoms having quadrupolar nuclei (^{2}H, ^{14}N, ^{81,79}Br, ^{35,37}Cl or ^{127}I). These nuclei are generally objects of NQR nuclear quadrupole resonance and NMR nuclear magnetic resonance studies. Interactions between the nuclear quadrupole moments Q and the electric field gradients (EFG) are characterized via the nuclear quadrupole coupling constants

$$QCC = \frac{e^2 qzzQ}{h}$$

where the eq_{zz} term corresponds to the principal component of the EFG tensor and e is the elementary charge. In turn, the electric field gradient opens the way to description of bonding modes in molecules when the QCC values are accurately determined by NMR or NQR methods.

In general, when ionic bonding occurs in the solid (or liquid) state, it is not possible to talk about a single "ionic bond" between two individual atoms, because the cohesive forces that keep the lattice together are of a more collective nature. This is quite different in the case of covalent bonding, where we can often speak of a distinct bond localized between two particular atoms. However, even if ionic bonding is combined with some covalency, the result is *not* necessarily discrete bonds of a localized character. In such cases, the resulting bonding often requires description in terms of a band structure consisting of gigantic molecular orbitals spanning the entire crystal. Thus, the bonding in the solid often retains its collective rather than localized nature. When the difference in electronegativity is decreased, the bonding may then lead to a semiconductor, a semimetal or eventually a metallic conductor with metallic bonding.

Comparison of Ionic and Covalent Bonds

A molecule or compound is made when two or more atoms form a chemical bond that links them together. As we have seen, there are two types of bonds: ionic bonds and covalent bonds. In an ionic bond, the atoms are bound together by the electrostatic forces in the attraction between ions of opposite charge. Ionic bonds usually occur between metal and nonmetal ions. For example, sodium (Na), a metal, and chloride (Cl), a nonmetal, form an ionic bond to make NaCl. In a covalent bond, the atoms bond by sharing electrons. Covalent bonds usually occur between nonmetals. For example, in water (H_2O) each hydrogen (H) and oxygen (O) share a pair of electrons to make a molecule of two hydrogen atoms single bonded to a single oxygen atom.

In general, ionic bonds occur between elements that are far apart on the periodic table. Covalent bonds occur between elements that are close together on the periodic table. Ionic compounds tend to be brittle in their solid form and have very high melting temperatures. Covalent compounds tend to be soft, and have relatively low melting and boiling points. Water, a liquid composed of covalently bonded molecules, can also be used as a test substance for other ionic and covalently compounds. Ionic compounds tend to dissolve in water (e.g., sodium chloride, NaCl); covalent compounds sometimes dissolve well in water (e.g., hydrogen chloride, HCl), and sometimes do not (e.g., butane, C_4H_{10}). Properties of ionic and covalent compounds are listed in Table.

Properties of Ionic and Covalent Compounds

Property	Ionic	Covalent
How bond is made	Transfer of e^-	Sharing of e^-
Bond is between	Metals and nonmetals	Nonmetals
Position on periodic table	Opposite sides	Close together
Dissolve in water?	Yes	Varies
Consistency	Brittle	Soft
Melting temperature	High	Low

The properties listed in Table are exemplified by sodium chloride (NaCl) and chlorine gas (Cl_2). Like other ionic compounds, sodium chloride contains a metal ion (sodium) and a nonmetal ion (chloride), is brittle, and has a high melting temperature. Chlorine gas is similar to other covalent compounds in that it is a nonmetal and has a very low melting temperature.

Sodium chloride (NaCl), an ionic compound.

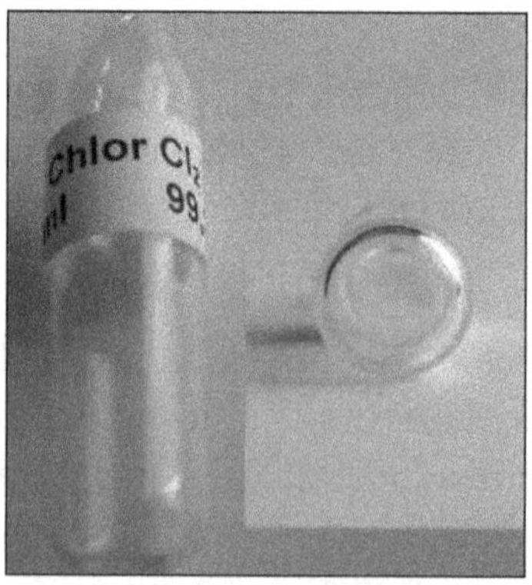

Chlorine gas (Cl_2), a covalent compound.

Metallic Bonding

Metallic bonding is a type of chemical bonding that rises from the electrostatic attractive force between conduction electrons (in the form of an electron cloud of delocalized electrons) and positively charged metal ions. It may be described as the sharing of *free* electrons among a structure of positively charged ions (cations). Metallic bonding accounts for many physical properties of metals, such as strength, ductility, thermal and electrical resistivity and conductivity, opacity, and luster.

Metallic bonding is not the only type of chemical bonding a metal can exhibit, even as a pure substance. For example, elemental gallium consists of covalently-bound pairs of atoms in both liquid and solid state—these pairs form a crystal structure with metallic bonding between them. Another example of a metal–metal covalent bond is mercurous ion (Hg_2^{2+}).

The Nature of Metallic Bonding

The combination of two phenomena gives rise to metallic bonding: Delocalization of electrons and the availability of a far larger number of delocalized energy states than of delocalized electrons. The latter could be called electron deficiency.

In 2D

Graphene is an example of two-dimensional metallic bonding. Its metallic bonds are similar to aromatic bonding in benzene, naphthalene, anthracene, ovalene, and so on.

In 3D

Metal aromaticity in metal clusters is another example of delocalization, this time often in three-dimensional entities. Metals take the delocalization principle to its extreme and one could say that a crystal of a metal represents a single molecule over which all conduction electrons are delocalized in all three dimensions. This means that inside the metal one can generally not distinguish molecules, so that the metallic bonding is neither intra- nor intermolecular. 'Nonmolecular' would perhaps be a better term. Metallic bonding is mostly non-polar, because even in alloys there is little difference among the electronegativities of the atoms participating in the bonding interaction (and, in pure elemental metals, none at all). Thus, metallic bonding is an extremely delocalized communal form of covalent bonding. In a sense, metallic bonding is not a 'new' type of bonding at all, therefore, and it describes the bonding only as present in a *chunk* of condensed matter, be it crystalline solid, liquid, or even glass. Metallic vapors by contrast are often atomic (Hg) or at times contain molecules like Na_2 held together by a more conventional covalent bond. This is why it is not correct to speak of a single 'metallic bond'.

The delocalization is most pronounced for s- and p-electrons. For caesium it is so strong that the electrons are virtually free from the caesium atoms to form a gas constrained only by the surface of the metal. For caesium, therefore, the picture of Cs^+ ions held together by a negatively charged electron gas is not too inaccurate. For other elements the electrons are less free, in that they still experience the potential of the metal atoms, sometimes quite strongly. They require a more intricate quantum mechanical treatment (e.g., tight binding) in which the atoms are viewed as neutral, much like the carbon atoms in benzene. For d- and especially f-electrons the delocalization is not strong at all and this explains why these electrons are able to continue behaving as unpaired electrons that retain their spin, adding interesting magnetic properties to these metals.

Electron Deficiency and Mobility

Metal atoms contain few electrons in their valence shells relative to their periods or energy levels. They are electron deficient elements and the communal sharing does not change that. There remain far more available energy states than there are shared electrons. Both requirements for conductivity are therefore fulfilled: strong delocalization and partly filled energy bands. Such electrons can therefore easily change from one energy state into a slightly different one. Thus, not only do they become delocalized, forming a sea of electrons permeating the structure, but they are also able to migrate through the structure when an external electrical field is imposed, leading to electrical conductivity. Without the field, there are electrons moving equally in all directions. Under the field, some will adjust their state slightly, adopting a different wave vector. As a consequence, there will be more moving one way than the other and a net current will result.

The freedom of conduction electrons to migrate also give metal atoms, or layers of them, the capacity to slide past each other. Locally, bonds can easily be broken and replaced by new ones after the deformation. This process does not affect the communal metallic bonding very much. This gives rise to metals' typical characteristic phenomena of malleability and ductility. This is particularly true for pure elements. In the presence of dissolved impurities, the defects in the structure that function as cleavage points may get blocked and the material becomes harder. Gold, for example, is very soft in pure form (24-karat), which is why alloys of 18-karat or lower are preferred in jewelry.

Metals are typically also good conductors of heat, but the conduction electrons only contribute partly to this phenomenon. Collective (i.e., delocalized) vibrations of the atoms known as phonons that travel through the solid as a wave, contribute strongly.

However, the latter also holds for a substance like diamond. It conducts heat quite well but *not* electricity. The latter is *not* a consequence of the fact that delocalization is absent in diamond, but simply that carbon is not electron deficient. The electron deficiency is an important point in distinguishing metallic from more conventional covalent bonding. Thus, we should amend the expression given above into: Metallic bonding is an extremely delocalized communal form of electron deficient covalent bonding.

Metallic Radius

Metallic radius is defined as one-half of the distance between the two adjacent metal ions in the metallic structure. This radius depends on the nature of the atom as well as its environment—specifically, on the coordination number (CN), which in turn depends on the temperature and applied pressure.

When comparing periodic trends in the size of atoms it is often desirable to apply so-called Goldschmidt correction, which converts the radii to the values the atoms would have if they were 12-coordinated. Since metallic radii are always biggest for the highest coordination number, correction for less dense coordinations involves multiplying by x, where $0 < x < 1$. Specifically, for CN = 4, x = 0.88; for CN = 6, x = 0.96, and for CN = 8, x = 0.97. The correction is named after Victor Goldschmidt who obtained the numerical values quoted above.

The radii follow general periodic trends: they decrease across the period due to increase in the effective nuclear charge, which is not offset by the increased number of valence electrons. The radii also increase down the group due to increase in principal quantum number. Between rows 3 and 4, the lanthanide contraction is observed – there is very little increase of the radius down the group due to the presence of poorly shielding f orbitals.

Strength of the Bond

The atoms in metals have a strong attractive force between them. Much energy is required to overcome it. Therefore, metals often have high boiling points, with tungsten (5828 K) being extremely high. A remarkable exception is the elements of the zinc group: Zn, Cd, and Hg. Their electron configuration ends in ns^2 and this comes to resemble a noble gas configuration like that of helium more and more when going down in the periodic table because the energy distance to the empty np orbitals becomes larger. These metals are therefore relatively volatile, and are avoided in ultra-high vacuum systems.

Otherwise, metallic bonding can be very strong, even in molten metals, such as Gallium. Even though gallium will melt from the heat of one's hand just above room temperature, its boiling point is not far from that of copper. Molten gallium is, therefore, a very nonvolatile liquid thanks to its strong metallic bonding.

The strong bonding of metals in the liquid form demonstrates that the energy of a metallic bond is not a strong function of the direction of the metallic bond; this lack of bond directionality is a direct consequence of electron delocalization, and is best understood in contrast to the directional bonding of covalent bonds. The energy of a metallic bond is thus mostly a function of the number of electrons which surround the metallic atom, as exemplified by the Embedded atom model. This typically results in metals assuming relatively simple, close-packed crystal structures, such as FCC, BCC, and HCP.

Given high enough cooling rates and appropriate alloy composition, metallic bonding can occur even in glasses with an amorphous structure.

Much biochemistry is mediated by the weak interaction of metal ions and biomolecules. Such interactions and their associated conformational change has been measured using dual polarisation interferometry.

Solubility and Compound Formation

Metals are insoluble in water or organic solvents unless they undergo a reaction with them. Typically this is an oxidation reaction that robs the metal atoms of their itinerant electrons, destroying the metallic bonding. However metals are often readily soluble in each other while retaining the metallic character of their bonding. Gold, for example, dissolves easily in mercury, even at room temperature. Even in solid metals, the solubility can be extensive. If the structures of the two metals are the same, there can even be complete solid solubility, as in the case of electrum, the alloys of silver and gold. At times, however, two metals will form alloys with different structures than either of the two parents. One could call these materials metal compounds, but, because materials with metallic bonding are typically not molecular, Dalton's law of integral proportions is not valid and often a range of stoichiometric ratios can be achieved. It is better to abandon such concepts as 'pure substance' or 'solute' is such cases and speak of phases instead. The study of such phases has traditionally been more the domain of metallurgy than of chemistry, although the two fields overlap considerably.

Localization and Clustering: From Bonding to Bonds

The metallic bonding in complicated compounds does not necessarily involve all constituent elements equally. It is quite possible to have an element or more that do not partake at all. One could picture the conduction electrons flowing around them like a river around an island or a big rock. It is possible to observe which elements do partake, e.g., by looking at the core levels in an X-ray photoelectron spectroscopy (XPS) spectrum. If an element partakes, its peaks tend to be skewed.

Some intermetallic materials e.g. do exhibit metal clusters, reminiscent of molecules and these compounds are more a topic of chemistry than of metallurgy. The formation of the clusters could be seen as a way to 'condense out' (localize) the electron deficient bonding into bonds of a more localized nature. Hydrogen is an extreme example of this form of condensation. At high pressures it is a metal. The core of the planet Jupiter could be said to be held together by a combination of metallic bonding and high pressure induced by gravity. At lower pressures however the bonding becomes entirely localized into a regular covalent bond. The localization is so complete that the (more familiar) H_2 gas results. A similar argument holds for an element like boron. Though it is electron deficient compared to carbon, it does not form a metal. Instead it has a number of complicated structures in which icosahedral B_{12} clusters dominate. Charge density waves are a related phenomenon.

As these phenomena involve the movement of the atoms towards or away from each other, they can be interpreted as the coupling between the electronic and the vibrational states (i.e. the phonons) of the material. A different such electron-phonon interaction is thought to cause a very different result at low temperatures, that of superconductivity. Rather than blocking the mobility of the charge carriers by forming electron pairs in localized bonds, Cooper-pairs are formed that no longer experience any resistance to their mobility.

Optical Properties

The presence of an ocean of mobile charge carriers has profound effects on the optical properties of metals. They can only be understood by considering the electrons as a *collective* rather than considering the states of individual electrons involved in more conventional covalent bonds.

Light consists of a combination of an electrical and a magnetic field. The electrical field is usually able to excite an elastic response from the electrons involved in the metallic bonding. The result is that photons are not able to penetrate very far into the metal and are typically reflected. They bounce off, although some may also be absorbed. This holds equally for all photons of the visible spectrum, which is why metals are often silvery white or grayish with the characteristic specular reflection of metallic luster. The balance between reflection and absorption determines how white or how gray they are, although surface tarnish can obscure such observations. Silver, a very good metal with high conductivity is one of the whitest.

Notable exceptions are reddish copper and yellowish gold. The reason for their color is that there is an upper limit to the frequency of the light that metallic electrons can readily respond to, the plasmon frequency. At the plasmon frequency, the frequency-dependent dielectric function of the free electron gas goes from negative (reflecting) to positive (transmitting); higher frequency photons are not reflected at the surface, and do not contribute to the color of the metal. There are some materials like indium tin oxide (ITO) that are metallic conductors (actually degenerate semiconductors) for which this threshold is in the infrared, which is why they are transparent in the visible, but good mirrors in the IR.

For silver the limiting frequency is in the far UV, but for copper and gold it is closer to the visible. This explains the colors of these two metals. At the surface of a metal resonance effects known as surface plasmons can result. They are collective oscillations of the conduction electrons like a ripple in the electronic ocean. However, even if photons have enough energy they usually do not have enough momentum to set the ripple in motion. Therefore, plasmons are hard to excite on a bulk metal. This is why gold and copper still look like lustrous metals albeit with a dash of color. However, in colloidal gold the metallic bonding is confined to a tiny metallic particle, preventing the oscillation wave of the plasmon from 'running away'. The momentum selection rule is therefore broken, and the plasmon resonance causes an extremely intense absorption in the green with a resulting beautiful purple-red color. Such colors are orders of magnitude more intense than ordinary absorptions seen in dyes and the like that involve individual electrons and their energy states.

Intermolecular Force

Intermolecular forces (IMF) are the forces which mediate interaction between molecules, including forces of attraction or repulsion which act between molecules and other types of neighboring particles, e.g. atoms or ions. Intermolecular forces are weak relative to intramolecular forces – the forces which hold a molecule together. For example, the covalent bond, involving sharing electron pairs between atoms, is much stronger than the forces present between neighboring molecules. Both sets of forces are essential parts of force fields frequently used in molecular mechanics.

The investigation of intermolecular forces starts from macroscopic observations which indicate the existence and action of forces at a molecular level. These observations include non-ideal-gas thermodynamic behavior reflected by virial coefficients, vapor pressure, viscosity, superficial tension, and absorption data.

The first reference to the nature of microscopic forces is found in Alexis Clairaut's work *Theorie de la Figure de la Terre*. Other scientists who have contributed to the investigation of microscopic forces include: Laplace, Gauss, Maxwell and Boltzmann.

Attractive intermolecular forces are categorized into the following types:

- Hydrogen bonding.
- Ionic bonding.
- Ion–induced dipole forces.
- Ion–dipole forces.
- Van der Waals forces – Keesom force, Debye force, and London dispersion force.

Information on intermolecular forces is obtained by macroscopic measurements of properties like viscosity, pressure, volume, temperature (PVT) data. The link to microscopic aspects is given by virial coefficients and Lennard-Jones potentials.

Dipole–dipole and Similar Interactions

Regular Dipole

Dipole–dipole interactions are electrostatic interactions between molecules which have permanent dipoles. This interaction is stronger than the London forces but is weaker than ion-ion interaction because only partial charges are involved. These interactions tend to align the molecules to increase attraction (reducing potential energy). An example of a dipole–dipole interaction can be seen in hydrogen chloride (HCl): The positive end of a polar molecule will attract the negative end of the other molecule and influence its position. Polar molecules have a net attraction between them. Examples of polar molecules include hydrogen chloride (HCl) and chloroform ($CHCl_3$).

$$\overset{\delta+}{H}-\overset{\delta-}{Cl}\cdots\overset{\delta+}{H}-\overset{\delta-}{Cl}$$

Often molecules contain dipolar groups of atoms, but have no overall dipole moment on the molecule as a whole. This occurs if there is symmetry within the molecule that causes the dipoles to cancel each other out. This occurs in molecules such as tetrachloromethane and carbon dioxide. The dipole–dipole interaction between two individual atoms is usually zero, since atoms rarely carry a permanent dipole. These forces are discussed further in the section about the Keesom interaction, below.

Ion–dipole and Ion–induced Dipole Forces

Ion–dipole and ion–induced dipole forces are similar to dipole–dipole and dipole–induced dipole interactions but involve ions, instead of only polar and non-polar molecules. Ion–dipole and ion–induced dipole forces are stronger than dipole–dipole interactions because the charge of any ion is much greater than the charge of a dipole moment. Ion–dipole bonding is stronger than hydrogen bonding.

An ion–dipole force consists of an ion and a polar molecule interacting. They align so that the positive and negative groups are next to one another, allowing maximum attraction. An important example of this interaction is hydration of ions in water which give rise to hydration enthalpy. The polar water molecules surround themselves around ions is water and the energy released during the process is known as hydration enthalpy. The interaction has its immense importance in justifying the stability of various ions (like Cu+2) in water.

An ion–induced dipole force consists of an ion and a non-polar molecule interacting. Like a dipole–induced dipole force, the charge of the ion causes distortion of the electron cloud on the non-polar molecule.

Relative Strength of Forces

Bond type	Dissociation energy (kcal/mol)	Dissociation energy (kJ/mol)	Note
Ionic lattice	250–4000	1100-20000	
Covalent bond	30–260	130–1100	
Hydrogen bond	1–12	4–50	About 5 kcal/mol (21 kJ/mol) in water

Dipole–dipole	0.5–2	2–8	
London dispersion forces	<1 to 15	<4 to 63	Estimated from the enthalpies of vaporization of hydrocarbons

This comparison is approximate. The actual relative strengths will vary depending on the molecules involved. Ionic bonding and covalent bonding will always be stronger than intermolecular forces in any given substance.

Effect on the Behavior of Gases

Intermolecular forces are repulsive at short distances and attractive at long distances. In a gas, the repulsive force chiefly has the effect of keeping two molecules from occupying the same volume. This gives a real gas a tendency to occupy a larger volume than an ideal gas at the same temperature and pressure. The attractive force draws molecules closer together and gives a real gas a tendency to occupy a smaller volume than an ideal gas. Which interaction is more important depends on temperature and pressure.

In a gas, the distances between molecules are generally large, so intermolecular forces have only a small effect. The attractive force is not overcome by the repulsive force, but by the thermal energy of the molecules. Temperature is the measure of thermal energy, so increasing temperature reduces the influence of the attractive force. In contrast, the influence of the repulsive force is essentially unaffected by temperature.

When a gas is compressed to increase its density, the influence of the attractive force increases. If the gas is made sufficiently dense, the attractions can become large enough to overcome the tendency of thermal motion to cause the molecules to disperse. Then the gas can condense to form a solid or liquid, i.e., a condensed phase. Lower temperature favors the formation of a condensed phase. In a condensed phase, there is very nearly a balance between the attractive and repulsive forces.

Quantum Mechanical Theories

Intermolecular forces observed between atoms and molecules can be described phenomenologically as occurring between permanent and instantaneous dipoles, as outlined above. Alternatively, one may seek a fundamental, unifying theory that is able to explain the various types of interactions such as hydrogen bonding, van der Waals forces and dipole–dipole interactions. Typically, this is done by applying the ideas of quantum mechanics to molecules, and Rayleigh–Schrödinger perturbation theory has been especially effective in this regard. When applied to existing quantum chemistry methods, such a quantum mechanical explanation of intermolecular interactions provides an array of approximate methods that can be used to analyze intermolecular interactions. One of the most helpful methods to visualize this kind of intermolecular interactions, that we can find in quantum chemistry, is the non-covalent interaction index, which is based on the electron density of the system. London Dispersion Forces (LDF) play a big role with this.

Hydrogen Bonding

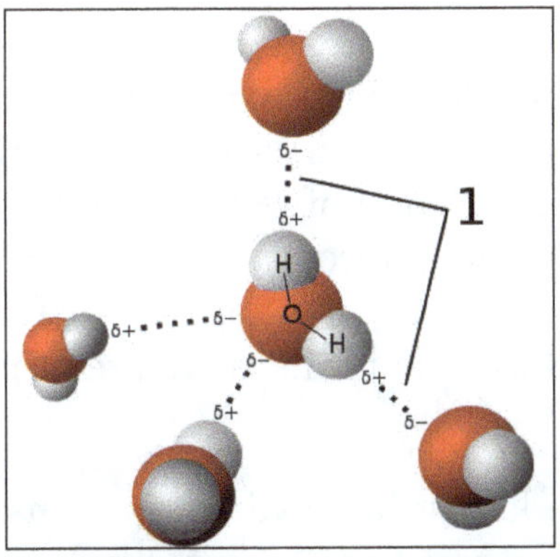

Model of hydrogen bonds (1) between molecules of water.

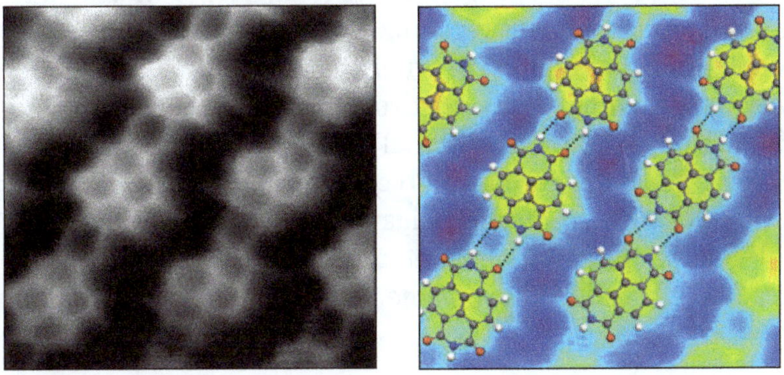

AFM image of naphthalenetetracarboxylic diimide molecules on silver-terminated silicon, interacting via hydrogen bonding, taken at 77 K. ("Hydrogen bonds" in the top image are exaggerated by artifacts of the imaging technique.)

A hydrogen bond (often informally abbreviated H-bond) is a partial intermolecular bonding interaction between a lone pair on an electron rich donor atom, particularly the second-row elements nitrogen (N), oxygen (O), or fluorine (F), and the antibonding orbital of a bond between hydrogen (H) and a more electronegative atom or group. Such an interacting system is generally denoted Dn–H···Ac, where the solid line denotes a polar covalent bond, and the dotted or dashed line indicates the hydrogen bond. The use of three centered dots for the hydrogen bond is specifically recommended by the IUPAC. While hydrogen bonding has both covelant and electrostatic contributions, and the degrees to which they contribute are currently debated, the present evidence strongly implies that the primary contribution is covalent.

Hydrogen bonds can be intermolecular (occurring between separate molecules) or intramolecular (occurring among parts of the same molecule). Depending on the nature of the donor and acceptor atoms which constitute the bond, their geometry, and environment, the energy of a hydrogen bond can vary between 1 and 40 kcal/mol. This makes them somewhat stronger than a van der Waals interaction, and weaker than fully covalent or ionic bonds. This type of bond can occur in inorganic molecules such as water and in organic molecules like DNA and proteins.

The hydrogen bond is responsible for many of the anomalous physical and chemical properties of compounds of N, O, and F. In particular, intermolecular hydrogen bonding is responsible for the high boiling point of water (100 °C) compared to the other group 16 hydrides that have much weaker hydrogen bonds. Intramolecular hydrogen bonding is partly responsible for the secondary and tertiary structures of proteins and nucleic acids. It also plays an important role in the structure of polymers, both synthetic and natural.

Weaker hydrogen bonds are known for hydrogen atoms bound to elements such as sulfur (S) or chlorine (Cl); even carbon (C) can serve as a donor, particularly when the carbon or one of its neighbors is electronegative (e.g., in chloroform, aldehydes and terminal acetylenes). Gradually, it was recognized that there are many examples of weaker hydrogen bonding involving donor Dn other than N, O, or F and/or acceptor Ac with electronegativity approaching that of hydrogen (rather than being much more electronegative). Though these "non-traditional" hydrogen bonding interactions are often quite weak (~1 kcal/mol), they are also ubiquitous and are increasingly recognized as important control elements in receptor-ligand interactions in medicinal chemistry or intra /intermolecular interactions in materials sciences. The definition of hydrogen bonding has gradually broadened over time to include these weaker attractive interactions.

As part of a more detailed list of criteria, the IUPAC publication acknowledges that the attractive interaction can arise from some combination of electrostatics (multipole-multipole and multipole-induced multipole interactions), covalency (charge transfer by orbital overlap), and dispersion (London forces), and states that the relative importance of each will vary depending on the system. However, a footnote to the criterion recommends the exclusion of interactions in which dispersion is the primary contributor, specifically giving Ar---CH_4 and CH_4---CH_4 as examples of such interactions to be excluded from the definition.

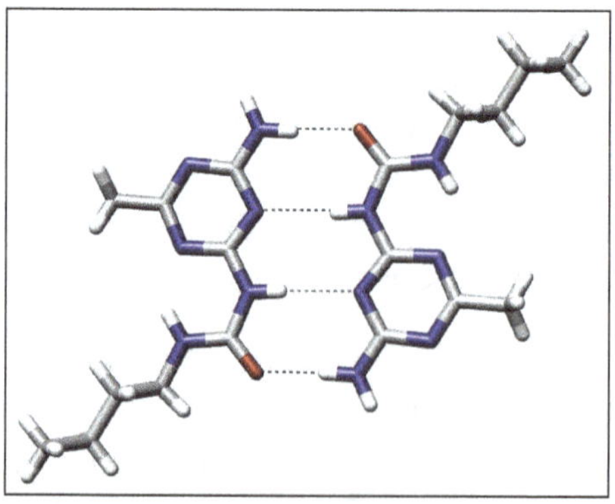

An example of intermolecular hydrogen bonding in a self-assembled dimer complex. The hydrogen bonds are represented by dotted lines.

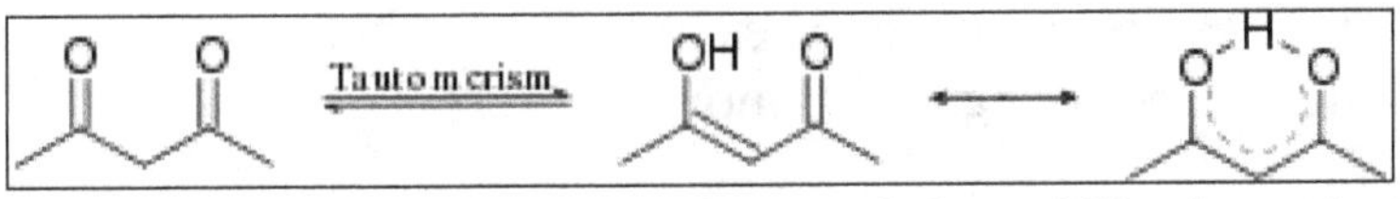

Intramolecular hydrogen bonding in acetylacetone helps stabilize the enol tautomer.

Definitions and General Characteristics

A hydrogen atom attached to a relatively electronegative atom is the hydrogen bond *donor*. C-H bonds only participate in hydrogen bonding when the carbon atom is bound to electronegative substituents, as is the case in chloroform, $CHCl_3$. In a hydrogen bond, the electronegative atom not covalently attached to the hydrogen is named proton acceptor, whereas the one covalently bound to the hydrogen is named the proton donor. While this nomenclature is recommended by the IUPAC, it can be misleading, since in other donor-acceptor bonds, the donor/acceptor assignment is based on the source of the electron pair (such nomenclature is also used for hydrogen bonds by some authors). In the hydrogen bond donor, the H center is protic. The donor is a Lewis acid. Hydrogen bonds are represented as H···Y system, where the dots represent the hydrogen bond. Liquids that display hydrogen bonding (such as water) are called associated liquids.

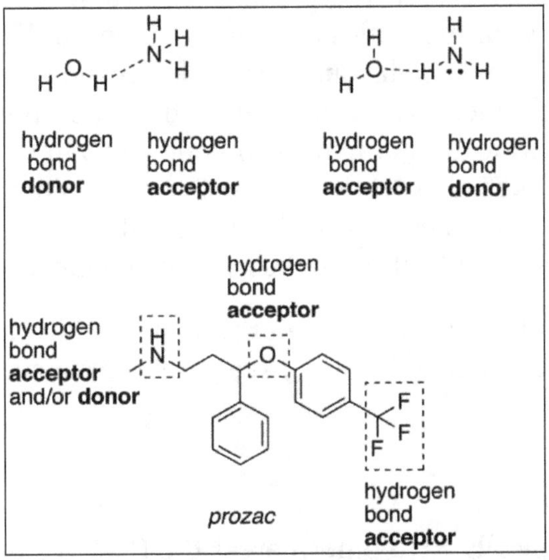

Examples of hydrogen bond donating (donors) and hydrogen bond accepting groups (acceptors).

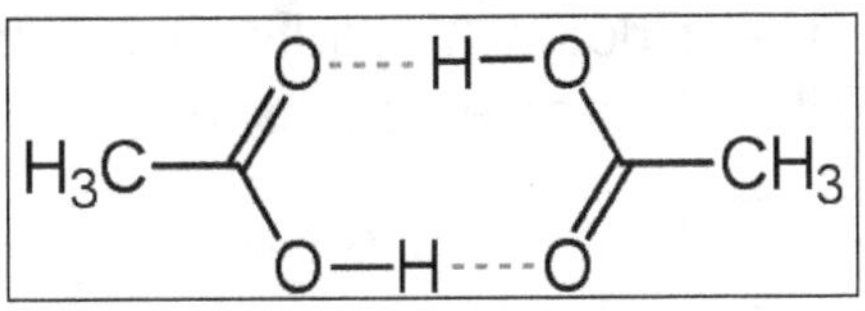

Cyclic dimer of acetic acid; dashed green lines represent hydrogen bonds.

The hydrogen bond is often described as an electrostatic dipole-dipole interaction. However, it also has some features of covalent bonding: it is directional and strong, produces interatomic distances shorter than the sum of the van der Waals radii, and usually involves a limited number of interaction partners, which can be interpreted as a type of valence. These covalent features are more substantial when acceptors bind hydrogens from more electronegative donors.

Bond Strength

Hydrogen bonds can vary in strength from weak (1–2 kJ mol^{-1}) to strong (161.5 kJ mol^{-1} in the ion HF^{-}_{2}). Typical enthalpies in vapor include:

- F–H···:F (161.5 kJ/mol or 38.6 kcal/mol), illustrated uniquely by HF_2^-, bifluoride.
- O–H···:N (29 kJ/mol or 6.9 kcal/mol), illustrated water-ammonia.
- O–H···:O (21 kJ/mol or 5.0 kcal/mol), illustrated water-water, alcohol-alcohol.
- N–H···:N (13 kJ/mol or 3.1 kcal/mol), illustrated by ammonia-ammonia.
- N–H···:O (8 kJ/mol or 1.9 kcal/mol), illustrated water-amide.
- HO–H···: OH_3^- (18 kJ/mol or 4.3 kcal/mol).

The strength of intermolecular hydrogen bonds is most often evaluated by measurements of equilibria between molecules containing donor and/or acceptor units, most often in solution. The strength of intramolecular hydrogen bonds can be studied with equilibria between conformers with and without hydrogen bonds. The most important method for the identification of hydrogen bonds also in complicated molecules is crystallography, sometimes also NMR-spectroscopy. Structural details, in particular distances between donor and acceptor which are smaller than the sum of the van der Waals radii can be taken as indication of the hydrogen bond strength.

One scheme gives the following somewhat arbitrary classification: those that are 15 to 40 kcal/mol, 5 to 15 kcal/mol, and >0 to 5 kcal/mol are considered strong, moderate, and weak, respectively.

Structural Details

The X–H distance is typically ≈110 pm, whereas the H···Y distance is ≈160 to 200 pm. The typical length of a hydrogen bond in water is 197 pm. The ideal bond angle depends on the nature of the hydrogen bond donor. The following hydrogen bond angles between a hydrofluoric acid donor and various acceptors have been determined experimentally:

Acceptor···donor	Vsepr Geometry	Angle (°)
HCN···HF	Linear	180
H_2CO···HF	Trigonal Planar	120
H_2O···HF	Pyramidal	46
H_2S···HF	Pyramidal	89
SO_2···HF	Trigonal	142

Spectroscopy

Strong hydrogen bonds are revealed by downfield shifts in the ^{1}H NMR spectrum. For example, the acidic proton in the enol tautomer of acetylacetone appears at δ_H 15.5, which is about 10 ppm downfield of a conventional alcohol.

In the IR spectrum, hydrogen bonding shifts the X-H stretching frequency to lower energy (i.e. the vibration frequency decreases). This shift reflects a weakening of the X-H bond. Certain hydrogen bonds - improper hydrogen bonds - show a blue shift of the X-H stretching frequency and a decrease in the bond length.

Theoretical Considerations

Hydrogen bonding is of continuing theoretical interest. According to a modern description O:H-O integrates both the intermolecular O:H lone pair ":" nonbond and the intramolecular H-O polar-covalent bond associated with O-O repulsive coupling.

Quantum chemical calculations of the relevant interresidue potential constants (compliance constants) revealed large differences between individual H bonds of the same type. For example, the central interresidue N−H···N hydrogen bond between guanine and cytosine is much stronger in comparison to the N−H···N bond between the adenine-thymine pair.

Theoretically, the bond strength of the hydrogen bonds can be assessed using NCI index, non-covalent interactions index, which allows a visualization of these non-covalent interactions, as its name indicases, using the electron density of the system.

From interpretations of the anisotropies in the Compton profile of ordinary ice that the hydrogen bond is partly covalent. However, this interpretation was challenged.

Most generally, the hydrogen bond can be viewed as a metric-dependent electrostatic scalar field between two or more intermolecular bonds. This is slightly different from the intramolecular bound states of, for example, covalent or ionic bonds; however, hydrogen bonding is generally still a bound state phenomenon, since the interaction energy has a net negative sum. The initial theory of hydrogen bonding proposed by Linus Pauling suggested that the hydrogen bonds had a partial covalent nature. This interpretation remained controversial until NMR techniques demonstrated information transfer between hydrogen-bonded nuclei, a feat that would only be possible if the hydrogen bond contained some covalent character.

Hydrogen Bonds in Small Molecules

Water

A ubiquitous example of a hydrogen bond is found between water molecules. In a discrete water molecule, there are two hydrogen atoms and one oxygen atom. Two molecules of water can form a hydrogen bond between them that is to say oxygen-hydrogen bonding; the simplest case, when only two molecules are present, is called the water dimer and is often used as a model system. When more molecules are present, as is the case with liquid water, more bonds are possible because the oxygen of one water molecule has two lone pairs of electrons, each of which can form a hydrogen bond with a hydrogen on another water molecule. This can repeat such that every water molecule is H-bonded with up to four other molecules, as shown in the figure (two through its two lone pairs, and two through its two hydrogen atoms). Hydrogen bonding strongly affects the crystal structure of ice, helping to create an open hexagonal lattice. The density of ice is less than the density of water at the same temperature; thus, the solid phase of water floats on the liquid, unlike most other substances.

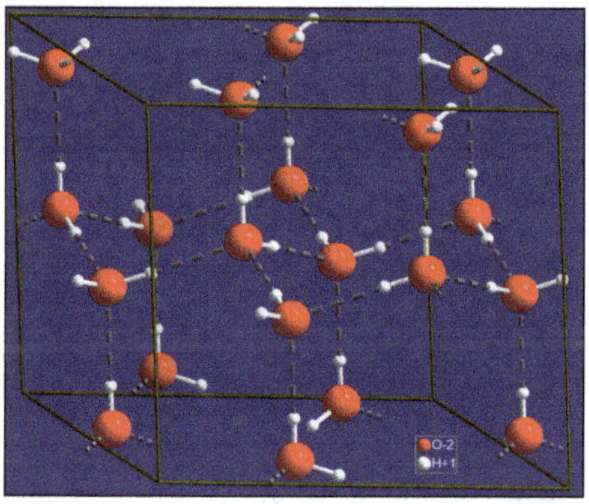

Crystal structure of hexagonal ice. Gray dashed lines indicate hydrogen bonds.

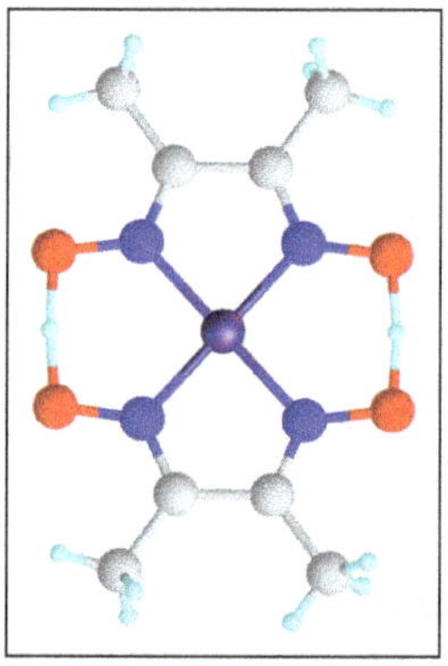

Structure of nickel bis(dimethylglyoximate), which features two linear hydrogen-bonds.

Liquid water's high boiling point is due to the high number of hydrogen bonds each molecule can form, relative to its low molecular mass. Owing to the difficulty of breaking these bonds, water has a very high boiling point, melting point, and viscosity compared to otherwise similar liquids not conjoined by hydrogen bonds. Water is unique because its oxygen atom has two lone pairs and two hydrogen atoms, meaning that the total number of bonds of a water molecule is up to four.

The number of hydrogen bonds formed by a molecule of liquid water fluctuates with time and temperature. From TIP4P liquid water simulations at 25 °C, it was estimated that each water molecule participates in an average of 3.59 hydrogen bonds. At 100 °C, this number decreases to 3.24 due to the increased molecular motion and decreased density, while at 0 °C, the average number of hydrogen bonds increases to 3.69. A more recent study found a much smaller number of hydrogen bonds: 2.357 at 25 °C. The differences may be due to the use of a different method for defining and counting the hydrogen bonds.

Where the bond strengths are more equivalent, one might instead find the atoms of two interacting water molecules partitioned into two polyatomic ions of opposite charge, specifically hydroxide (OH^-) and hydronium (H_3O^+). (Hydronium ions are also known as "hydroxonium" ions.)

$$H{-}O^- \ H_3O^+$$

Indeed, in pure water under conditions of standard temperature and pressure, this latter formulation is applicable only rarely; on average about one in every 5.5×10^8 molecules gives up a proton to another water molecule, in accordance with the value of the dissociation constant for water under such conditions. It is a crucial part of the uniqueness of water.

Because water may form hydrogen bonds with solute proton donors and acceptors, it may competitively inhibit the formation of solute intermolecular or intramolecular hydrogen bonds. Consequently, hydrogen bonds between or within solute molecules dissolved in water are almost always unfavorable relative to hydrogen bonds between water and the donors and acceptors for hydrogen bonds on those solutes. Hydrogen bonds between water molecules have an average lifetime of 10^{-11} seconds, or 10 picoseconds.

Bifurcated and Over-coordinated Hydrogen Bonds in Water

A single hydrogen atom can participate in two hydrogen bonds, rather than one. This type of bonding is called "bifurcated" (split in two or "two-forked"). It can exist, for instance, in complex natural or synthetic organic molecules. It has been suggested that a bifurcated hydrogen atom is an essential step in water reorientation. Acceptor-type hydrogen bonds (terminating on an oxygen's lone pairs) are more likely to form bifurcation (it is called over-coordinated oxygen, OCO) than are donor-type hydrogen bonds, beginning on the same oxygen's hydrogens.

Other Liquids

For example, hydrogen fluoride—which has three lone pairs on the F atom but only one H atom—can form only two bonds; (ammonia has the opposite problem: three hydrogen atoms but only one lone pair).

$$H{-}F{\cdots}H{-}F{\cdots}H{-}F$$

Further Manifestations of Solvent Hydrogen Bonding

- Increase in the melting point, boiling point, solubility, and viscosity of many compounds can be explained by the concept of hydrogen bonding.
- Negative azeotropy of mixtures of HF and water.
- The fact that ice is less dense than liquid water is due to a crystal structure stabilized by hydrogen bonds.
- Dramatically higher boiling points of NH_3, H_2O, and HF compared to the heavier analogues PH_3, H_2S, and HCl, where hydrogen-bonding is absent.
- Viscosity of anhydrous phosphoric acid and of glycerol.
- Dimer formation in carboxylic acids and hexamer formation in hydrogen fluoride, which occur even in the gas phase, resulting in gross deviations from the ideal gas law.
- Pentamer formation of water and alcohols in apolar solvents.

Hydrogen Bonds in Polymers

Hydrogen bonding plays an important role in determining the three-dimensional structures and the properties adopted by many synthetic and natural proteins.

DNA

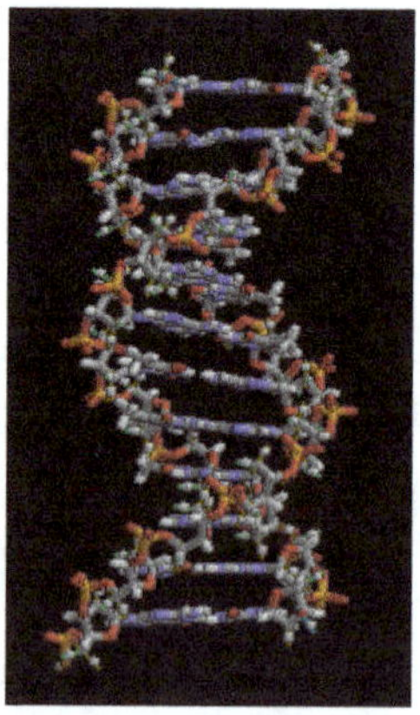

The structure of part of a DNA double helix.

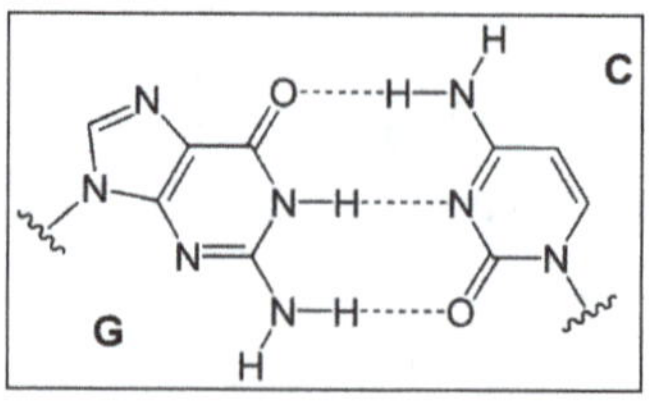

Hydrogen bonding between guanine and cytosine, one of two types of base pairs in DNA.

In these macromolecules, bonding between parts of the same macromolecule cause it to fold into a specific shape, which helps determine the molecule's physiological or biochemical role. For example, the double helical structure of DNA is due largely to hydrogen bonding between its base pairs (as well as pi stacking interactions), which link one complementary strand to the other and enable replication.

Proteins

In the secondary structure of proteins, hydrogen bonds form between the backbone oxygens and amide hydrogens. When the spacing of the amino acid residues participating in a hydrogen bond occurs regularly between positions i and $i + 4$, an alpha helix is formed. When the spacing is less, between positions i and $i + 3$, then a 3_{10} helix is formed. When two strands are joined by hydrogen bonds involving alternating residues on each participating strand, a beta sheet is formed. Hydrogen bonds also play a part in forming the tertiary structure of protein through interaction of R-groups.

Bifurcated H-bond systems are common in alpha-helical transmembrane proteins between the backbone amide C=O of residue i as the H-bond acceptor and two H-bond donors from residue $i+4$: the backbone amide N-H and a side-chain hydroxyl or thiol H^+. The energy preference of the bifurcated H-bond hydroxyl or thiol system is -3.4 kcal/mol or -2.6 kcal/mol, respectively. This type of bifurcated H-bond provides an intrahelical H-bonding partner for polar side-chains, such as serine, threonine, and cysteine within the hydrophobic membrane environments.

The role of hydrogen bonds in protein folding has also been linked to osmolyte-induced protein stabilization. Protective osmolytes, such as trehalose and sorbitol, shift the protein folding equilibrium toward the folded state, in a concentration dependent manner. While the prevalent explanation for osmolyte action relies on excluded volume effects that are entropic in nature, recent circular dichroism (CD) experiments have shown osmolyte to act through an enthalpic effect. The molecular mechanism for their role in protein stabilization is still not well established, though several mechanisms have been proposed. Recently, computer molecular dynamics simulations suggested that osmolytes stabilize proteins by modifying the hydrogen bonds in the protein hydration layer.

Several studies have shown that hydrogen bonds play an important role for the stability between subunits in multimeric proteins. For example, a study of sorbitol dehydrogenase displayed an important hydrogen bonding network which stabilizes the tetrameric quaternary structure within the mammalian sorbitol dehydrogenase protein family.

A protein backbone hydrogen bond incompletely shielded from water attack is a dehydron. Dehydrons promote the removal of water through proteins or ligand binding. The exogenous dehydration enhances the electrostatic interaction between the amide and carbonyl groups by de-shielding their partial charges. Furthermore, the dehydration stabilizes the hydrogen bond by destabilizing the nonbonded state consisting of dehydrated isolated charges.

Wool, being a protein fibre, is held together by hydrogen bonds, causing wool to recoil when stretched. However, washing at high temperatures can permanently break the hydrogen bonds and a garment may permanently lose its shape.

Cellulose

Hydrogen bonds are important in the structure of cellulose and derived polymers in its many different forms in nature, such as cotton and flax.

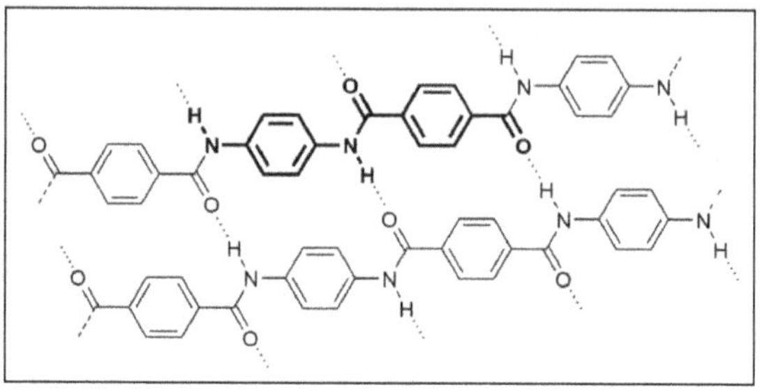

Para-aramid structure.

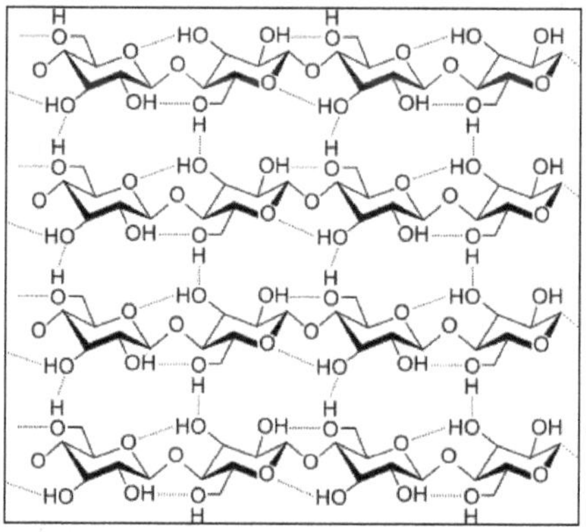

A strand of cellulose (conformation I_α), showing the hydrogen bonds (dashed) within and between cellulose molecules.

Synthetic Polymers

Many polymers are strengthened by hydrogen bonds within and between the chains. Among the synthetic polymers, a well characterized example is nylon, where hydrogen bonds occur in the repeat unit and play a major role in crystallization of the material. The bonds occur between carbonyl and amine groups in the amide repeat unit. They effectively link adjacent chains, which help reinforce the material. The effect is great in aramid fibre, where hydrogen bonds stabilize the linear chains laterally. The chain axes are aligned along the fibre axis, making the fibres extremely stiff and strong.

The hydrogen-bond networks make both natural and synthetic polymers sensitive to humidity levels in the atmosphere because water molecules can diffuse into the surface and disrupt the network. Some polymers are more sensitive than others. Thus nylons are more sensitive than aramids, and nylon 6 more sensitive than nylon-11.

Symmetric Hydrogen Bond

A symmetric hydrogen bond is a special type of hydrogen bond in which the proton is spaced exactly halfway between two identical atoms. The strength of the bond to each of those atoms is equal. It is an example of a three-center four-electron bond. This type of bond is much stronger than a "normal" hydrogen bond. The effective bond order is 0.5, so its strength is comparable to a covalent bond. It is seen in ice at high pressure, and also in the solid phase of many anhydrous acids such as hydrofluoric acid and formic acid at high pressure. It is also seen in the bifluoride ion $[F\text{--}H\text{--}F]^-$. Due to severe steric constraint, the protonated form of Proton Sponge (1,8-bis(dimethylamino) naphthalene) and its derivatives also have symmetric hydrogen bonds ($[N\text{--}H\text{--}N]^+$), although in the case of protonated Proton Sponge, the assembly is bent.

Symmetric hydrogen bonds have been observed recently spectroscopically in formic acid at high pressure (>GPa). Each hydrogen atom forms a partial covalent bond with two atoms rather than one. Symmetric hydrogen bonds have been postulated in ice at high pressure (Ice X). Low-barrier hydrogen bonds form when the distance between two heteroatoms is very small.

Dihydrogen Bond

The hydrogen bond can be compared with the closely related dihydrogen bond, which is also an intermolecular bonding interaction involving hydrogen atoms. These structures have been known for some time, and well characterized by crystallography; however, an understanding of their relationship to the conventional hydrogen bond, ionic bond, and covalent bond remains unclear. Generally, the hydrogen bond is characterized by a proton acceptor that is a lone pair of electrons in nonmetallic atoms (most notably in the nitrogen, and chalcogen groups). In some cases, these proton acceptors may be pi-bonds or metal complexes. In the dihydrogen bond, however, a metal hydride serves as a proton acceptor, thus forming a hydrogen-hydrogen interaction. Neutron diffraction has shown that the molecular geometry of these complexes is similar to hydrogen bonds, in that the bond length is very adaptable to the metal complex/hydrogen donor system.

Dynamics Probed by Spectroscopic Means

The dynamics of hydrogen bond structures in water can be probed by the IR spectrum of OH stretching vibration. In the hydrogen bonding network in protic organic ionic plastic crystals (POIPCs), which are a type of phase change material exhibiting solid-solid phase transitions prior to melting, variable-temperature infrared spectroscopy can reveal the temperature dependence of hydrogen bonds and the dynamics of both the anions and the cations. The sudden weakening of hydrogen bonds during the solid-solid phase transition seems to be coupled with the onset of orientational or rotational disorder of the ions.

Application to Drugs

Hydrogen bonding is a key to the design of drugs. According to Lipinski's rule of five the majority of orally active drug tend to have between five and ten hydrogen bonds. These interactions exist between nitrogen–hydrogen and oxygen–hydrogen centers. As with many other rules of thumb, many exceptions exist.

Hydrogen Bonding Phenomena

- Occurrence of proton tunneling during DNA replication is believed to be responsible for cell mutations.
- High water solubility of many compounds such as ammonia is explained by hydrogen bonding with water molecules.
- Deliquescence of NaOH is caused in part by reaction of OH^- with moisture to form hydrogen-bonded H_3O^- species. An analogous process happens between $NaNH_2$ and NH_3, and between NaF and HF.
- The presence of hydrogen bonds can cause an anomaly in the normal succession of states of matter for certain mixtures of chemical compounds as temperature increases or decreases. These compounds can be liquid until a certain temperature, then solid even as the temperature increases, and finally liquid again as the temperature rises over the "anomaly interval".
- Smart rubber utilizes hydrogen bonding as its sole means of bonding, so that it can "heal" when torn, because hydrogen bonding can occur on the fly between two surfaces of the same polymer.

Van Der Waals Force

Van der Waals forces help geckos walk effortlessly along walls and ceilings.

Rain water flux from a canopy. Among the forces that govern drop formation: van der Waals force, surface tension, cohesion, Plateau–Rayleigh instability.

In molecular physics, the van der Waals force, named after Dutch scientist Johannes Diderik van der Waals, is a distance-dependent interaction between atoms or molecules. Unlike ionic or covalent bonds, these attractions do not result from a chemical electronic bond; they are comparatively weak and therefore more susceptible to disturbance. The van der Waals force quickly vanishes at longer distances between interacting molecules.

Van der Waals force plays a fundamental role in fields as diverse as supramolecular chemistry, structural biology, polymer science, nanotechnology, surface science, and condensed matter physics. It also underlies many properties of organic compounds and molecular solids, including their solubility in polar and non-polar media.

If no other force is present, the distance between atoms at which the force becomes repulsive rather than attractive as the atoms approach one another is called the van der Waals contact distance; this phenomenon results from the mutual repulsion between the atoms' electron clouds. The van der Waals force has the same origin as the Casimir effect, arising from quantum interactions with the zero-point field.

The term *van der Waals force* is sometimes used loosely for all intermolecular forces. The term always includes the London dispersion force between instantaneously induced dipoles. It is sometimes applied to the Debye force between a permanent dipole and a corresponding induced dipole or to the Keesom force between permanent molecular dipoles.

Van der Waals forces include attraction and repulsions between atoms, molecules, and surfaces, as well as other intermolecular forces. They differ from covalent and ionic bonding in that they are caused by correlations in the fluctuating polarizations of nearby particles (a consequence of quantum dynamics).

Being the weakest of the weak chemical forces, with a strength between 0.4 and 4kJ/mol, they may still support an integral structural load when multitudes of such interactions are present.

The force results from a transient shift in electron density. Specifically, the electron density may temporarily shift more greatly to one side of the nucleus. This generates a transient charge to which a nearby atom can be either attracted or repelled. When the interatomic distance of two atoms is greater than 0.6 nm the force is not strong enough to be observed. In the same vein, when the interatomic distance is below 0.4 nm the force becomes repulsive.

Intermolecular forces have four major contributions:

- A repulsive component resulting from the Pauli exclusion principle that prevents the collapse of molecules.
- Attractive or repulsive electrostatic interactions between permanent charges (in the case of molecular ions), dipoles (in the case of molecules without inversion center), quadrupoles (all molecules with symmetry lower than cubic), and in general between permanent multipoles. The electrostatic interaction is sometimes called the Keesom interaction or Keesom force after Willem Hendrik Keesom.
- Induction (also known as polarization), which is the attractive interaction between a permanent multipole on one molecule with an induced multipole on another. This interaction is sometimes called Debye force after Peter J.W. Debye.
- Dispersion (usually named London dispersion interactions after Fritz London), which is the attractive interaction between any pair of molecules, including non-polar atoms, arising from the interactions of instantaneous multipoles.

All intermolecular/van der Waals forces are anisotropic (except those between two noble gas atoms), which means that they depend on the relative orientation of the molecules. The induction and dispersion interactions are always attractive, irrespective of orientation, but the electrostatic interaction changes sign upon rotation of the molecules. That is, the electrostatic force can be attractive or repulsive, depending on the mutual orientation of the molecules. When molecules are in thermal motion, as they are in the gas and liquid phase, the electrostatic force is averaged out to a large extent, because the molecules thermally rotate and thus probe both repulsive and attractive parts of the electrostatic force. Sometimes this effect is expressed by the statement that "random thermal motion around room temperature can usually overcome or disrupt them" (which refers to the electrostatic component of the van der Waals force). Clearly, the thermal averaging effect is much less pronounced for the attractive induction and dispersion forces.

The Lennard-Jones potential is often used as an approximate model for the isotropic part of a total (repulsion plus attraction) van der Waals force as a function of distance.

Van der Waals forces are responsible for certain cases of pressure broadening (van der Waals broadening) of spectral lines and the formation of van der Waals molecules. The

London-van der Waals forces are related to the Casimir effect for dielectric media, the former being the microscopic description of the latter bulk property. The first detailed calculations of this were done in 1955 by E. M. Lifshitz. A more general theory of van der Waals forces has also been developed.

The main characteristics of van der Waals forces are:

- They are weaker than normal covalent and ionic bonds.
- Van der Waals forces are additive and cannot be saturated.
- They have no directional characteristic.
- They are all short-range forces and hence only interactions between the nearest particles need to be considered (instead of all the particles). Van der Waals attraction is greater if the molecules are closer.
- Van der Waals forces are independent of temperature except dipole – dipole interactions.

In low molecular weight alcohols, the hydrogen-bonding properties of their polar hydroxyl group dominate other weaker van der Waals interactions. In higher molecular weight alcohols, the properties of the nonpolar hydrocarbon chain(s) dominate and determine their solubility.

Van Der Waals Forces between Macroscopic Objects

For macroscopic bodies with known volumes and numbers of atoms or molecules per unit volume, the total van der Waals force is often computed based on the "microscopic theory" as the sum over all interacting pairs. It is necessary to integrate over the total volume of the object, which makes the calculation dependent on the objects' shapes. For example, the van der Waals' interaction energy between spherical bodies of radii R_1 and R_2 and with smooth surfaces was approximated in 1937 by Hamaker (using London's famous 1937 equation for the dispersion interaction energy between atoms/molecules as the starting point) by:

$$U(z;R_1,R_2)=-\frac{A}{6}\left(\frac{2R_1R_2}{z^2-(R_1+R_2)^2}+\frac{2R_1R_2}{z^2-(R_1-R_2)^2}+\ln\left[\frac{z^2-(R_1+R_2)^2}{z^2-(R_1-R_2)^2}\right]\right)$$

where A is the Hamaker coefficient, which is a constant (~10^{-19} – 10^{-20} J) that depends on the material properties (it can be positive or negative in sign depending on the intervening medium), and z is the center-to-center distance; i.e., the sum of R_1, R_2, and r (the distance between the surfaces):

$$z=R_1+R_2+r$$

In the limit of close-approach, the spheres are sufficiently large compared to the distance between them; i.e., $r \ll R_1$ or R_2, so that equation,

$$U(z;R_1,R_2) = -\frac{A}{6}\left(\frac{2R_1R_2}{z^2-(R_1+R_2)^2}+\frac{2R_1R_2}{z^2-(R_1-R_2)^2}+\ln\left[\frac{z^2-(R_1+R_2)^2}{z^2-(R_1-R_2)^2}\right]\right)$$

for the potential energy function simplifies to:

$$U(r;R_1,R_2) = -\frac{AR_1R_2}{(R_1+R_2)6r}$$

The van der Waals force between two spheres of constant radii (R1 and R2 are treated as parameters) is then a function of separation since the force on an object is the negative of the derivative of the potential energy function, $F_{VW}(r) = -\frac{d}{dr}U(r)$. This yields:

$$F_{VW}(r) = -\frac{AR_1R_2}{(R_1+R_2)6r^2}$$

From the expression above, it is seen that the van der Waals force decreases with decreasing size of bodies (R). Nevertheless, the strength of inertial forces, such as gravity and drag/lift, decrease to a greater extent. Consequently, the van der Waals forces become dominant for collections of very small particles such as very fine-grained dry powders (where there are no capillary forces present) even though the force of attraction is smaller in magnitude than it is for larger particles of the same substance. Such powders are said to be cohesive, meaning they are not as easily fluidized or pneumatically conveyed as their more coarse-grained counterparts. Generally, free-flow occurs with particles greater than about 250 μm.

The van der Waals force of adhesion is also dependent on the surface topography. If there are surface asperities, or protuberances, that result in a greater total area of contact between two particles or between a particle and a wall, this increases the van der Waals force of attraction as well as the tendency for mechanical interlocking.

The microscopic theory assumes pairwise additivity. It neglects many-body interactions and retardation. A more rigorous approach accounting for these effects, called the "macroscopic theory" was developed by Lifshitz in 1956. Langbein derived a much more cumbersome "exact" expression in 1970 for spherical bodies within the framework of the Lifshitz theory while a simpler macroscopic model approximation had been made by Derjaguin as early as 1934. Expressions for the van der Waals forces for many different geometries using the Lifshitz theory have likewise been published.

Use by Geckos and Arthropods

The ability of geckos – which can hang on a glass surface using only one toe – to climb on sheer surfaces has been for many years mainly attributed to the van der Waals forces between these surfaces and the spatulae, or microscopic projections, which cover the hair-like setae found on their footpads. A later study suggested that capillary adhesion might play a role, but that hypothesis has been rejected by more recent studies.

Among the arthropods, some spiders have similar setae on their scopulae or scopula pads, enabling them to climb or hang upside-down from extremely smooth surfaces such as glass or porcelain.

Gecko climbing a glass surface.

London Dispersion Force

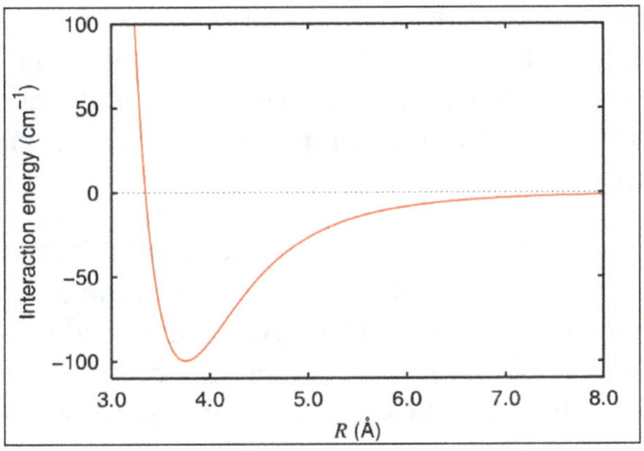

Interaction energy of argon dimer. The long-range section is due to London dispersion forces.

London dispersion forces (LDF, also known as dispersion forces, London forces, instantaneous dipole–induced dipole forces, or loosely van der Waals forces) are a type of force acting between atoms and molecules. They are part of the van der Waals forces. The LDF is named after the German physicist Fritz London.

The electron distribution around an atom or molecule undergoes fluctuations in time. These fluctuations create instantaneous electric fields which are felt by other nearby atoms and molecules, which in turn adjust the spatial distribution of their own electrons. The net effect is that the fluctuations in electron positions in one atom induce a corresponding redistribution of electrons in other atoms, such that the electron motions become correlated. While the detailed theory requires a quantum-mechanical explanation, the effect is frequently described as the formation of the *instantaneous dipoles* that (when separated by vacuum) attract each other. The magnitude of the London dispersion force is frequently described in terms of a single parameter called the Hamaker constant, typically symbolized A. For atoms that are located closer together than the wavelength of light, the interaction is essentially instantaneous and is described in terms of a "non-retarded" Hamaker constant. For entities that are farther apart, the finite time required for the fluctuation at one atom to be felt at a second atom ("retardation") requires use of a "retarded" Hamaker constant.

While the London dispersion force between individual atoms and molecules is quite weak and decreases quickly with separation (R) like $\frac{1}{R^6}$, in condensed matter (liquids and solids), the effect is cumulative over the volume of materials, or within and between organic molecules, such that London dispersion forces can be quite strong in bulk solid and liquids and decays much more slowly with distance. For example, the total force per unit area between two bulk solids decreases like — where R is the separation between them. The effects of London dispersion forces are most obvious in systems that are very non-polar (e.g., that lack ionic bonds), such as hydrocarbons and highly symmetric molecules such as bromine ($Br_{2,}$ a liquid at room temperature), iodine ($I_{2,}$ a solid at room temperature). In hydrocarbons and waxes the dispersion forces are sufficient to cause condensation from the gas phase into the liquid or solid phase. Sublimation heats of e.g. hydrocarbon crystals reflect the dispersion interaction. Liquification of oxygen and nitrogen gases into liquid phases is also dominated by attractive London dispersion forces.

When atoms/molecules are separated by a third medium(rather than vacuum), the situation becomes more complex. In aqueous solutions, the effects of dispersion forces between atoms/molecules are frequently less pronounced due to competition with polarizable solvent molecules. That is, the instantaneous fluctuations in one atom or molecule are felt both by the solvent (water) and by other molecules.

London forces become stronger as the atom in question becomes larger, and to a smaller degree for large molecules. This is due to the increased polarizability of molecules with larger, more dispersed electron clouds. The polarizability is a measure of ease with which electrons can be redistributed; a large polarizability implies that the electrons are more easily redistributed. This trend is exemplified by the halogens (from smallest to largest: F_2, Cl_2, Br_2, I_2). The same increase of dispersive attraction occurs within and between organic molecules in the order RF<RCL<RBr<RI, or with other more polarizable heteroatoms. Fluorine and chlorine are gases at room temperature, bromine is a liquid, and iodine is a solid. The London forces are thought to be arise from the motion of electrons.

Quantum Mechanical Theory of Dispersion Forces

The first explanation of the attraction between noble gas atoms was given by Fritz London in 1930. He used a quantum-mechanical theory based on second-order perturbation theory. The perturbation is because of the Coulomb interaction between the electrons and nuclei of the two moieties (atoms or molecules). The second-order perturbation expression of the interaction energy contains a sum over states. The states appearing in this sum are simple products of the stimulated electronic states of the monomers. Thus, no intermolecular antisymmetrization of the electronic states is included and the Pauli exclusion principle is only partially satisfied.

London wrote Taylor series expansion of the perturbation in $\frac{1}{R}$, where R is the distance between the nuclear centers of mass of the moieties.

This expansion is known as the multipole expansion because the terms in this series can be regarded as energies of two interacting multipoles, one on each monomer. Substitution of the multipole-expanded form of V into the second-order energy yields an expression that resembles somewhat an expression describing the interaction between instantaneous multipoles (see the qualitative description above). Additionally, an approximation, named after Albrecht Unsöld, must be introduced in order to obtain a description of London dispersion in terms of dipole polarizabilities and ionization potentials.

In this manner, the following approximation is obtained for the dispersion interaction E_{AB}^{disp} between two atoms A and B. Here α_A and α_B are the dipole polarizabilities of the respective atoms. The quantities I_A and I_B are the first ionization potentials of the atoms, and R is the intermolecular distance.

$$E_{AB}^{\text{disp}} \approx -\frac{3}{2}\frac{I_A I_B}{I_A + I_B}\frac{\alpha_A \alpha_B}{R^6}$$

This final London equation does not contain instantaneous dipoles. The "explanation" of the dispersion force as the interaction between two such dipoles was invented after London arrived at the proper quantum mechanical theory.

Relative Magnitude

Dispersion forces are usually dominant of the three van der Waals forces (orientation, induction, dispersion) between atoms and molecules, with the exception of molecules that are small and highly polar, such as water. The following contribution of the dispersion to the total intermolecular interaction energy has been given:

Contribution of the dispersion to the total intermolecular interaction energy	
Molecule pair	% of the total energy of interaction
Ne-Ne	100
CH_4-CH_4	100
HCl-HCl	86
HBr-HBr	96
HI-HI	99
CH_3Cl-CH_3Cl	68
NH_3-NH_3	57
H_2O-H_2O	24

HCl-HI	96
H_2O-CH_4	87

Causes of London Dispersion Forces

When you think of electrons around an atom, you probably picture tiny moving dots, spaced equally around the atomic nucleus. However, electrons are always in motion, and sometimes there are more on one side of an atom than on the other. This happens around any atom, but it's more pronounced in compounds because electrons feel the attractive pull of the protons of neighboring atoms. The electrons from two atoms can be arranged so that they produce temporary (instantaneous) electric dipoles. Even though the polarization is temporary, it's enough to affect the way atoms and molecules interact with each other. Through the inductive effect, or -I Effect, a permanent state of polarization occurs.

Dispersion forces occur between all atoms and molecules, regardless of whether they are polar or nonpolar. The forces come into play when the molecules are very close to each other. However, London dispersion forces are generally stronger between easily polarized molecules and weaker between molecules that are not easily polarized.

The magnitude of the force is related to the size of the molecule. Dispersion forces are stronger for larger and heavier atoms and molecules than for smaller and lighter ones. This is because the valence electrons are farther away from the nucleus in large atoms/molecules than in small ones, so they are not as tightly bound to the protons.

The shape or conformation of a molecule affects its polarizability. It's like fitting together blocks or playing Tetris, a video game—first introduced in 1984—that involves matching tiles. Some shapes will naturally line up better than others.

Consequences of London Dispersion Forces

The polarizability affects how easily atoms and molecules form bonds with each other, so it also affects properties such as melting point and boiling point. For example, if you consider Cl_2 (chlorine) and Br_2 (bromine), you might expect the two compounds to behave similarly because they are both halogens. Yet, chlorine is a gas at room temperature, while bromine is a liquid. This is because the London dispersion forces between the larger bromine atoms bring them close enough to form a liquid, while the smaller chlorine atoms have enough energy for the molecule to remain gaseous.

References

- Covalent-bond-examples, examples/science-480: softschools.com, Retrieved 16 March, 2020
- Polarization-of-Covalent-Bond, covalent-bond: byjus.com, Retrieved 03 May, 2020
- Chemical-bonding-and-molecular-structure/covalent-compounds, guides-chemistry: toppr.com, Retrieved 09 June, 2020

- Ionic Interactions in Natural and Synthetic Macromolecules, Editors: Alberto Ciferri, Angelo Perico, Wiley 2012, p 35-47; https://onlinelibrary.wiley.com/doi/abs/10.1002/9781118165850.ch2
- Brewer, Scott H.; Franzen, Stefan (2002). "Indium Tin Oxide Plasma Frequency Dependence on Sheet Resistance and Surface Adlayers Determined by Reflectance FTIR Spectroscopy". The Journal of Physical Chemistry B. 106 (50): 12986–12992. doi:10.1021/jp026600x
- Weinhold, Frank; Klein, Roger A. (2014). "What is a hydrogen bond? Resonance covalency in the supramolecular domain". Chemistry Education Research and Practice. 15: 276–285. doi:10.1039/c4rp00030g
- Klimchitskaya, G. L.; Mostepanenko, V. M. (July 2015). "Casimir and van der Waals Forces: Advances and Problems". Proceedings of Peter the Great St. Petersburg Polytechnic University (517): 41–65. arXiv:1507.02393. doi:10.5862/PROC.516.4
- Definition-of-london-dispersion-force-605313: thoughtco.com, Retrieved 28 July, 2020

4

Solid-State Synthesis

There are many processes and reactions that fall under solid-State synthesis. Some of them are sol-gel process, chemical transport reaction, solid-gas reaction, co-precipitation synthesis, hydrothermal synthesis, intercalation, etc. This chapter sheds light on the subject of solid-State synthesis for an in-depth understanding of it.

Solid-State Reaction

Solid-State Reaction is a chemical reaction between solids, the occurrence of which is detected by a characteristic color. Solid-state reactions also include reactions that result in the precipitation or dissolution of a colored deposit.

The procedure for carrying out analytical solid-state reactions is simple: Small amounts (of the order of 1 mg) of a reagent and of the substance to be tested are mixed on a strip of filter paper or in a porcelain crucible, and the resulting color is observed. This method can be used to detect, for example, Ni^{2+} in its salts; here, small amounts of dimethylglyoxime and $(NH_4)_2CO_3$ are added to the sample, as a result of which red dimethylglyoximine, $Ni(C_4H_7O_2N)_2$, is observed. Salts of Pb^{2+} react with KI to yield yellow PbI_2, and salts of Fe^{3+} react with $K_4Fe(CN)_6$ to give Prussian blue, $Fe_4[Fe(CN_6)]_3$.

Solid-State reactions can be used in the field to identify minerals, ores, and chemical fertilizers. They can also be used to test pharmaceuticals.

Solid-State reactions are performed in the absence of solvents by either grinding or melting the starting materials together or simply applying heat to a mixture of starting materials. This type of reaction is usually performed in order to obtain polycrystalline inorganic solids but may also be used in organic synthesis. Solid-State reactions offer reduced costs, decreased amounts of chemical waste and, sometimes, an increase in yield.

Solid-State reactions have been studied in atmospheres which extend from the residual gases present in high ($<10^{-3}$ Torr) or ultra-high ($<10^{-8}$ Torr) vacuum, through various pressures of selected inert, oxidising or reducing gases, to high pressure systems. Near ambient pressure, the atmosphere may be either static or constantly renewed in a flow system.

The presence of an inert atmosphere may alter the apparent kinetic characteristics of a rate process by hindering the removal of gaseous products or, in rapid exothermic reactions, by dissipating heat. When the prevailing atmosphere contains an appreciable pressure of a gaseous product of a reversible reaction (e.g. CO_2 from carbonate decomposition, NH_3 and H_2O from decomposition of ammonium salts or H_2O from dehydration reactions) then the kinetic behaviour shows a dependence on partial pressure. Garn has developed a technique whereby the partial pressure is generated by the reaction concerned. Even in a continuously evacuated system, there is a pressure gradient established between the reactant container and the system exit and it is difficult to measure this gradient. Solid products of metastable structure are formed in a number of reactions in vacuum. Gases may be used to suppress sublimation processes . The prevailing atmosphere may also exert some control on the rates of reactions which are essentially solidsolid interactions.

When a reactive atmosphere is present, the reaction undergone by a solid may be changed, and perhaps complicated, by the presence of a concurrent or consecutive gas—solid reaction. When one or more of the products of the reaction in vacuum is an oxidizable metal or lower oxide, the presence of oxygen will be particularly significant if the normal product and the oxidized product have a different reactivity for the interface process. A reducing atmosphere (hydrogen) accelerates the decomposition of nickel oxalate by removing product surface oxide and so enhancing the activity of the metal at the interface.

Reaction Mechanisms

Every chemical reaction proceeds according to a reaction mechanism, which is a step-by-step description of what occurs during a reaction on the molecular level. Each step of the mechanism is known as an elementary process, which describes a single moment during a reaction in which molecules break and/or form new bonds. It is important to keep in mind that every reaction mechanism is simply a proposed version of what might be occurring at the molecular level; even if a mechanism agrees with experimental results, it is impossible to prove a reaction mechanism for certain. That being said, there are two strict requirements that must be fulfilled for a reaction mechanism to be valid. They are:

- The sum of each elementary step in a reaction mechanism must yield the overall reaction equation.
- The rate law of the *rate-determining step* must agree with the experimentally determined rate law.

The rate-determining step is the slowest step in a reaction mechanism. Because it is the slowest, it determines the rate of the overall reaction.

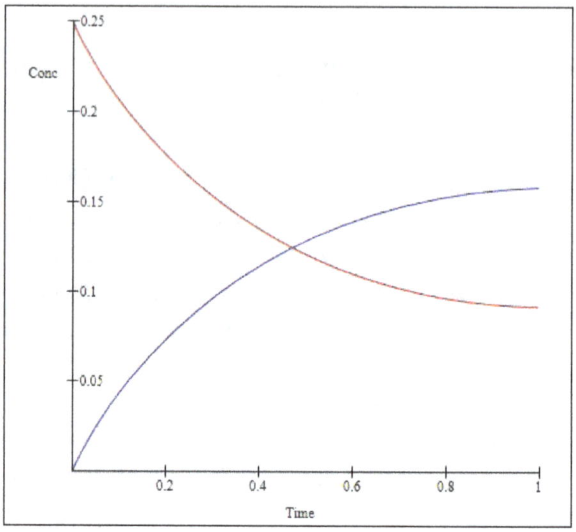

Change in concentration of chemicals over time: A plot of time versus concentration for two species in chemical equilibrium.

Elementary Reactions and their Rate Laws

Elementary reactions are classified in terms of their *molecularity*; that is, by how many molecules are involved in the given step. There are three possible molecularites:

- Unimolecular (A→products)
- Bimolecular (A+B→products)
- Termolecular (A+B+C→products)

Termolecularity is not common due to the rarity of three molecules colliding at the same time, and in just the right way for reaction to occur.

Example of Elementary Steps and their Rate Laws

Consider the following reaction:

$$2NO(g)+O_2(g)\rightarrow 2NO_2(g)$$

The overall equation suggests that two NO molecules collide with an oxygen molecule, forming NO_2. However, because termolecular reactions are extremely rare, this reaction most likely consists of two or more elementary steps. Upon further analysis, the mechanism is given as:

$$2NO(g)\rightarrow N_2O_2(g) \qquad \text{rate} = k_1[NO]^2$$

$$N_2O_2(g) + O_2(g) \rightarrow 2NO_2(g) \qquad rate = k_2[N_2O_2][O_2]$$

Note that the two steps here add to the overall reaction equation, as the intermediate N_2O_2 cancels. However, we cannot simply add the rate laws of each elementary step in order to get the overall reaction rate.

Rate-determining Steps

The rate of a multi-step reaction is determined by the slowest elementary step, which is known as the rate-determining step.

Chemists often write chemical equations for reactions as a single step that shows only the net result of a reaction. However, most chemical reactions occur over a series of elementary reactions. The complete sequence of these elementary steps is called a reaction mechanism. The reaction mechanism is the step-by-step process by which reactants actually become products. It is the "how" of the reaction, whereas the overall balanced equation shows only the "what" of the reaction. In kinetics, the rate of a reaction with several steps is determined by the slowest step, which is known as the rate-determining, or rate-limiting, step.

Rate Laws and the Rate-determining Step

Take the following example of a gas phase reaction:

$$CO + NO_2 \rightarrow CO_2 + NO$$

If this reaction occurred in a single step, its rate law would be:

$$r = k[NO_2][CO]$$

However, experiments show that the rate equation is:

$$r = k[NO_2]^2$$

The fact that the experimentally-determined rate law does not match the rate law derived from the overall reaction equation suggests that the reaction occurs over multiple steps. Further, the experimental rate law is second-order, suggesting that the reaction rate is determined by a step in which two NO_2 molecules react, and therefore the CO molecule must enter at another, faster step. A possible mechanism that explains the rate equation is:

$$2NO_2 \rightarrow NO_3 + NO \quad \text{(slow step, rate-determining)}$$

$$NO_3 + CO \rightarrow NO_2 + CO_2 \quad \text{(fast step)}$$

Since the first step is the slowest, and the entire reaction must wait for it, it is the rate-determining step. We can picture the rate-determining step to be like the narrowest point in an hourglass; it is the "bottleneck" point of the reaction that determines how quickly reactants can become products.

If the first step in a mechanism is rate-determining, it is easy to find the rate law for the overall expression from the mechanism. If the second or a later step is rate-determining, determining the rate law is slightly more complicated. We will explore how to write that rate law later.

Overall Reaction Rate Laws

Rate laws for reactions are affected by the position of the rate-determining step in the overall reaction mechanism.

Slow Step Followed by a Fast Step

If the first step in a reaction mechanism is the slow, rate-determining step, then the overall rate law for the reaction is easy to write, and simply follows the stoichiometry of the initial step. For example, consider the following reaction:

$$H_2(g)+2ICl(g) \rightarrow I_2(g)+2HCl(g)$$

The proposed reaction mechanism is given as follows:

$$\text{Slow reaction: } H_2+ICl \rightarrow HI+HCl \quad \text{rate}_1 = k_1[H_2][ICl]$$

$$\text{Fast reaction: } HI+ICl \rightarrow I_2+HCl \quad \text{rate}_2 = k_2[HI][ICl]$$

Since the first step is the rate-determining step, the overall reaction rate for this reaction is given by this step: rate=k[H_2][ICl]. As it turns out, this rate law has been verified with experimental evidence.

Fast Step Followed by a Slow Step

If the rate-determining step is not the first step in the reaction mechanism, the derivation of the rate law becomes slightly more complex. Consider the following reaction:

$$2NO(g)+O_2(g) \rightarrow 2NO_2(g)$$

The proposed mechanism is given by:

$$\text{Fast: } 2NO(g) \rightleftharpoons N_2O_2(g) \quad \text{rate}_1 = k_1[NO_2]$$

$$\text{Slow: } N_2O_2(g)+O_2(g) \rightarrow 2NO_2(g) \quad \text{rate}_2 = k_2[N_2O_2][O_2]$$

Step two is the slow, rate-determining step, so it might seem reasonable to assume that the rate law for this step should be the overall rate law for the reaction. However, this rate law contains N_2O_2, which is a reaction intermediate, and not a final product. The overall rate law cannot contain any such intermediates, because the rate law is determined by experiment only, and such intermediates are not observable. To get around this, we need to go back and consider the first step, which involves an equilibrium between NO and N_2O_2. At equilibrium, the rate of the forward reaction will equal the rate of the reverse reaction. We can write this as follows:

$$k_1[NO]^2 = k_{-1}[N_2O_2]$$

Rearranging for $[N_2O_2]$, we have:

$$[N_2O_2] = \frac{k_1}{k_{-1}}[NO]_2$$

We can now substitute this expression into the rate law for the second, rate-determining step. This yields the following:

$$\text{rate} = k_2\left(\frac{k_1}{k_{-1}}[NO]^2\right)[O_2] = k[NO]^2[O_2]$$

This overall rate law, which is second-order in NO and first-order in O_2, has been confirmed experimentally.

Rate law for a mechanism with a fast initial step: How to determine the rate law for a mechanism with a fast initial step. Remember, the overall rate law must be determined by experiment. Therefore, the rate law must contain no reaction intermediates.

Steady State Approximation

The steady state approximation can be used to determine the overall rate law when the rate-determining step is unknown.

We have assumed that every reaction proceeds according to a mechanism that is made up of elementary steps, and that there is always one elementary step in the mechanism that is the slowest. This slowest step determines the rate of the entire reaction, and as such, it is called the rate-determining step. We will now consider cases in which the rate-determining step is either unknown or when more than one step in the mechanism is slow, which affects the overall reaction rate. Both cases can be addressed by using what is known as the steady state approximation.

Applying the Steady State Approximation

Recall our mechanism for the reaction of nitric oxide and oxygen:

$$2\,NO(g) \rightleftharpoons N_2O_2(g)$$
$$N_2O_2(g) + O_2(g) \xrightarrow{k_2} 2\,NO_2(g)$$

Before, we assumed that the first step was fast, and that the second step was slow, thereby making it rate-determining. We will now proceed as if we had no such prior knowledge, and we do not know which, if either, of these steps is rate-determining. In such a case, we must assume that the reaction rate of each elementary step is equal, and the overall rate law for the reaction will be the final step in the mechanism, since this is the step that gives us our final products. In this case, the overall rate law will be:

$$\text{rate} = k_2[N_2O_2][O_2]$$

However, this rate law contains a reaction intermediate, which is not permitted in this process. We need to write this rate law in terms of reactants only. In order to do so, we must assume that the state of the reaction intermediate, N_2O_2, remains *steady*, or constant, throughout the course of the reaction. The idea is analogous to a tub being filled with water while the drain is open. At a certain point, the flow of water into the tub will equal the flow of water out of the tub, so that the height of the water in the tub remains constant. In reality, however, water is flowing into and out of the tub at all times; the overall amount of water in the tub at any given time does not change. Our reaction intermediate, N_2O_2, is like the water in the tub, because it is being produced and consumed at equal rates. Therefore, we can rewrite our initial equilibrium step as the following combination of reverse reactions:

$$2\,NO(g) \xrightarrow{k_1} N_2O_2(g)$$
$$2\,NO(g) \xleftarrow{k_{-1}} N_2O_2(g)$$

Here, k_1 and k_{-1} are the rate constants for the forward and reverse reactions, respectively. With the steady state assumption, we can write the following:

$$\text{rate of formation of } N_2O_2 = \text{rate of disappearance of } N_2O_2$$

Now, both of these rates can be written as rate laws derived from our elementary steps. We can write them as follows:

$$k_1[NO]^2 = k_{-1}[N_2O_2] + k_2[N_2O_2][O_2]$$

Note that the second side of the equation, which equals the rate of disappearance of our reaction intermediate N_2O_2, is written with two terms. The first term takes into account the disappearance of N_2O_2 in the reverse reaction of the initial step, and the second term takes into account the disappearance of N_2O_2 as a *reactant* in the second elementary step. We can rearrange this equation in terms of $[N_2O_2]$:

$$k_1[NO]^2 = [N_2O_2](k_{-1} + k_2[O_2])$$

$$\frac{k_1[NO]^2}{(k_{-1} + k_2[O_2])} = [N_2O_2]$$

Lastly, now that we have $[N_2O_2]$ in terms of our reactants only, we can substitute this entire expression into our equation for the overall rate law (equation $\text{rate} = k_2[N_2O_2][O_2]$ above). This yields the following:

$$\text{rate} = \frac{k_1k_2[NO]^2[O_2]}{k_{-1} + k_2[O_2]}$$

Here we have our final rate law for the overall reaction. We had no knowledge of the rate-determining step, so we used the steady state approximation for our reaction intermediate, N_2O_2.

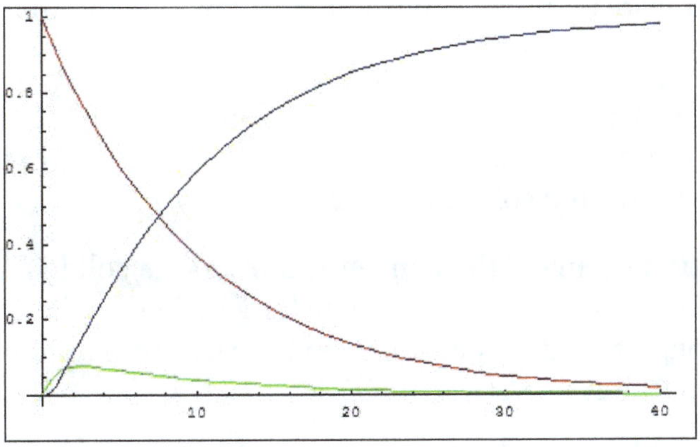

Concentration of reactants, products, and intermediates vs. time: Reactant concentrations are indicated in red, product concentrations are indicated in blue, and intermediate concentrations are indicated in green. Notice that after their initial production, the concentration of the reaction intermediates remains relatively constant (slope of green curve is approximately zero) throughout the course of the reaction.

Precipitation

Precipitation is the creation of a solid from a solution. When the reaction occurs in a liquid solution, the solid formed is called the 'precipitate'. The chemical that causes the solid to form is called the 'precipitant'. Without sufficient force of gravity (settling) to bring the solid particles together, the precipitate remains in suspension. After sedimentation, especially when using a centrifuge to press it into a compact mass, the precipitate may be referred to as a 'pellet'. Precipitation can be used as a medium. The precipitate-free liquid remaining above the solid is called the 'supernate' or 'supernatant'. Powders derived from precipitation have also historically been known as 'flowers'. When the solid appears in the form of cellulose fibers which have been through chemical processing, the process is often referred to as regeneration.

Chemical Precipitation.

Sometimes the formation of a precipitate indicates the occurrence of a chemical reaction. If silver nitrate solution is poured into a solution of sodium chloride, a chemical reaction occurs forming a white precipitate of silver chloride. When potassium iodide solution reacts with lead(II) nitrate solution, a yellow precipitate of lead(II) iodide is formed.

The precipitation may occur if concentration of a compound exceeds its solubility (such as when mixing solvents or changing their temperature). Precipitation may also occur rapidly from a supersaturated solution.

In solids, precipitation occurs if the concentration of one solid is above the solubility limit in the host solid, due to e.g. rapid quenching or ion implantation, and the temperature is high enough that diffusion can lead to segregation into precipitates. Precipitation in solids is routinely used to synthesize nanoclusters.

An important stage of the precipitation process is the onset of nucleation. The creation of a hypothetical solid particle includes the formation of an interface, which requires some energy based on the relative surface energy of the solid and the solution. If this energy is not available, and no suitable nucleation surface is available, supersaturation occurs.

Applications

Crystals of *meso*-tetratolylporphyrin from a reflux of propionic acid precipitate on cooling.

Copper from a wire is displaced by silver in a silver nitrate solution it is dipped into, and solid silver precipitates out.

Precipitation reactions can be used for making pigments, removing salts from water in water treatment, and in classical qualitative inorganic analysis.

Precipitation is also useful to isolate the products of a reaction during workup. Ideally, the product of the reaction is insoluble in the reaction solvent. Thus, it precipitates as it is formed, preferably forming pure crystals. An example of this would be the synthesis of porphyrins in refluxing propionic acid. By cooling the reaction mixture to room temperature, crystals of the porphyrin precipitate, and are collected by filtration.

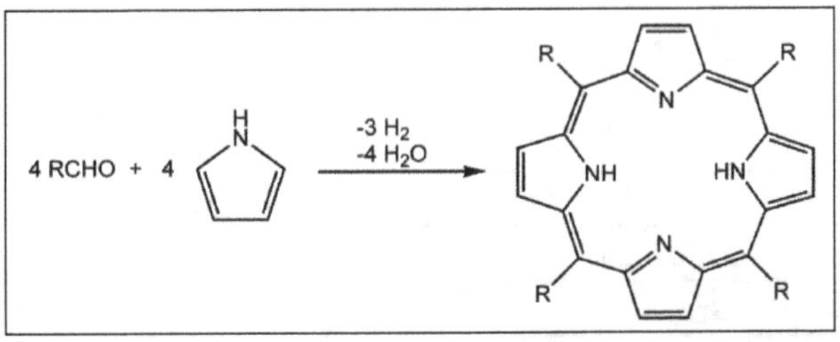

Precipitation may also occur when an *antisolvent* (a solvent in which the product is insoluble) is added, drastically reducing the solubility of the desired product. Thereafter, the precipitate may easily be separated by filtration, decanting, or centrifugation. An example would be the synthesis of chromic tetraphenylporphyrin chloride: water is added to the DMF reaction solution, and the product precipitates. Precipitation is also useful in purifying products: crude bmim-Cl is taken up in acetonitrile, and dropped into ethyl acetate, where it precipitates. Another important application of an antisolvent is in ethanol precipitation of DNA.

In metallurgy, precipitation from a solid solution is also a useful way to strengthen alloys; this process is known as solid solution strengthening.

Representation using Chemical Equations

An example of a precipitation reaction: Aqueous silver nitrate ($AgNO_3$) is added to a solution containing potassium chloride (KCl), the precipitation of a white solid, silver chloride (AgCl), is observed.

$$AgNO_3 + KCl \rightarrow AgCl\downarrow + KNO_3$$

The silver chloride (AgCl) has formed a solid, which is observed as a precipitate.

This reaction can be written emphasizing the dissociated ions in a combined solution. This is known as the ionic equation.

$$Ag^+ + NO_3^- + K^+ + Cl^- \rightarrow AgCl\downarrow + K^+ + NO_3^-$$

A final way to represent a precipitate reaction is known as a *net ionic reaction*.

Precipitate Colors

Many compounds containing metal ions produce precipitates with distinctive colors. The following are typical colors for various metals. However, many of these compounds can produce colors very different from those listed.

Gold	black
Chromium	Deep green, murky green, orange, purple, yellow, brown

Cobalt	Pink
Copper	Blue
Iron(II)	Green
Iron(III)	Reddish brown
Manganese	Pale pink
Nickel	Green
Lead	Yellow

Other compounds generally form white precipitates.

Green and reddish brown stains on a limestone core sample, respectively corresponding to precipitates of oxides/hydroxides of Fe^{2+} and Fe^{3+}.

Anion/Cation Analysis

Precipitate formation is useful in the detection of the type of cation in a salt. To do this, an alkali first reacts with the unknown salt to produce a precipitate that is the hydroxide of the unknown salt. To identify the cation, the color of the precipitate and its solubility in excess are noted. Similar processes are often used in sequence – for example, a barium nitrate solution will react with sulfate ions to form a solid barium sulfate precipitate, indicating that it is likely that sulfate ions are present.

Digestion

Digestion, or *precipitate ageing*, happens when a freshly formed precipitate is left, usually at a higher temperature, in the solution from which it precipitates. It results in cleaner and bigger particles. The physico-chemical process underlying digestion is called Ostwald ripening.

Co-precipitation

In chemistry, coprecipitation (CPT) or co-precipitation is the carrying down by a precipitate of substances normally soluble under the conditions employed. Analogously, in medicine, coprecipitation is specifically the precipitation of an unbound "antigen along with an antigen-antibody complex".

Coprecipitation is an important issue in chemical analysis, where it is often undesirable, but in some cases it can be exploited. In gravimetric analysis, which consists on precipitating the analyte and measuring its mass to determine its concentration or purity, coprecipitation is a problem because undesired impurities often coprecipitate with the analyte, resulting in excess mass. This problem can often be mitigated by "digestion" (waiting for the precipitate to equilibrate and form larger, purer particles) or by redissolving the sample and precipitating it again.

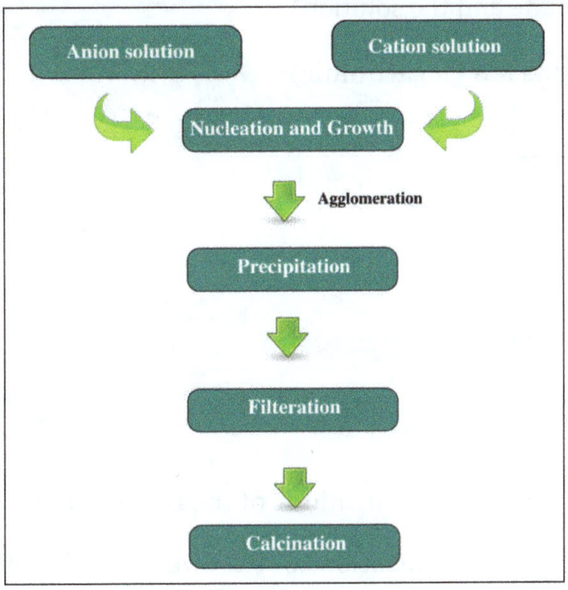

Typical co-precipitation method for micro and nano particle synthesis.

On the other hand, in the analysis of trace elements, as is often the case in radiochemistry, coprecipitation is often the only way of separating an element. Since the trace element is too dilute (sometimes less than a part per trillion) to precipitate by conventional means, it is typically coprecipitated with a *carrier*, a substance that has a similar crystalline structure that can incorporate the desired element. An example is the separation of francium from other radioactive elements by coprecipitating it with caesium salts such as caesium perchlorate. Otto Hahn is credited for promoting the use of coprecipitation in radiochemistry.

There are three main mechanisms of coprecipitation: inclusion, occlusion, and adsorption. An inclusion occurs when the impurity occupies a lattice site in the crystal structure of the carrier, resulting in a crystallographic defect; this can happen when the ionic radius and charge of the impurity are similar to those of the carrier. An adsorbate is an impurity that is weakly bound (adsorbed) to the surface of the precipitate. An occlusion occurs when an adsorbed impurity gets physically trapped inside the crystal as it grows.

Besides its applications in chemical analysis and in radiochemistry, coprecipitation is also "potentially important to many environmental issues closely related to water resources, including acid mine drainage, radionuclide migration in fouled waste repositories, metal contaminant transport at industrial and defense sites, metal concentrations in aquatic systems, and wastewater treatment technology". Coprecipitation is also used as a method of magnetic nanoparticle synthesis.

Distribution between Precipitate and Solution

There are two models describing of the distribution of the tracer compound between the two phases (the precipitate and the solution):

- Doerner-Hoskins law (logarithmic):

$$\ln\frac{a}{a-x} = \lambda \ln\frac{b}{b-y}$$

- Berthelot-Nernst law:

$$\frac{x}{a-x} = D\frac{y}{b-y}$$

where:

- a and b are the initial concentrations of the tracer and carrier, respectively;
- a-x and b-y are the concentrations of tracer and carrier after separation;
- x and y are the amounts of the tracer and carrier on the precipitate;
- D and λ are the distribution coefficients.

For D and λ greater than unity, the precipitate is enriched in the tracer.

Depending on the co-precipitation system and conditions either λ or D may be constant.

The derivation of the Doerner-Hoskins law assumes that there in no mass exchange between the interior of the precipitating crystals and the solution. When this assumption is fulfilled, then the content of the tracer in the crystal is non-uniform (the crystals are said to be heterogeneous). When the Berthelot-Nernst law applies, then the concentration of the tracer in the interior of the crystal is uniform (and the crystals are said to be homogeneous). This is the case when diffusion in the interior is possible (like in the liquids) or when the initial small crystals are allowed to recrystallize. Kinetic effects (like speed of crystallization and presence of mixing) play a role.

Sol-gel Process

In materials science, the sol–gel process is a method for producing solid materials from small molecules. The method is used for the fabrication of metal oxides, especially the oxides of silicon (Si) and titanium (Ti). The process involves conversion of monomers into a colloidal solution (*sol*) that acts as the precursor for an integrated network (or *gel*) of either discrete particles or network polymers. Typical precursors are metal alkoxides.

Stages in the Process

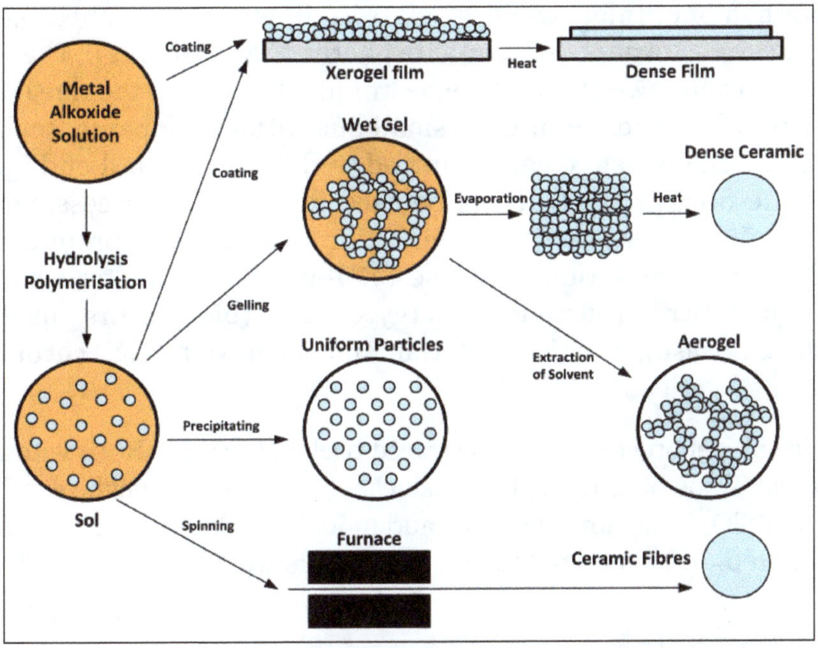

Schematic representation of the different stages and routes of the sol–gel technology.

In this chemical procedure, a "sol" (a colloidal solution) is formed that then gradually evolves towards the formation of a gel-like diphasic system containing both a liquid phase and solid phase whose morphologies range from discrete particles to continuous polymer networks. In the case of the colloid, the volume fraction of particles (or particle density) may be so low that a significant amount of fluid may need to be removed initially for the gel-like properties to be recognized. This can be accomplished in any number of ways. The simplest method is to allow time for sedimentation to occur, and then pour off the remaining liquid. Centrifugation can also be used to accelerate the process of phase separation.

Removal of the remaining liquid (solvent) phase requires a drying process, which is typically accompanied by a significant amount of shrinkage and densification. The rate at which the solvent can be removed is ultimately determined by the distribution of porosity in the gel. The ultimate microstructure of the final component will clearly be strongly influenced by changes imposed upon the structural template during this phase of processing.

Afterwards, a thermal treatment, or firing process, is often necessary in order to favor further polycondensation and enhance mechanical properties and structural stability via final sintering, densification, and grain growth. One of the distinct advantages of using this methodology as opposed to the more traditional processing techniques is that densification is often achieved at a much lower temperature.

The precursor sol can be either deposited on a substrate to form a film (e.g., by dip-coating or spin coating), cast into a suitable container with the desired shape (e.g., to obtain monolithic ceramics, glasses, fibers, membranes, aerogels), or used to synthesize powders (e.g., microspheres, nanospheres). The sol–gel approach is a cheap and low-temperature technique that allows the fine control of the product's chemical composition. Even small quantities of dopants, such as organic dyes and rare-earth elements, can be introduced in the sol and end up uniformly dispersed in the final product. It can be used in ceramics processing and manufacturing as an investment casting material, or as a means of producing very thin films of metal oxides for various purposes. Sol–gel derived materials have diverse applications in optics, electronics, energy, space, (bio)sensors, medicine (e.g., controlled drug release), reactive material, and separation (e.g., chromatography) technology.

The interest in sol–gel processing can be traced back in the mid-1800s with the observation that the hydrolysis of tetraethyl orthosilicate (TEOS) under acidic conditions led to the formation of SiO_2 in the form of fibers and monoliths. Sol–gel research grew to be so important that in the 1990s more than 35,000 papers were published worldwide on the process.

Particles and Polymers

The sol–gel process is a wet-chemical technique used for the fabrication of both glassy and ceramic materials. In this process, the sol (or solution) evolves gradually towards the formation of a gel-like network containing both a liquid phase and a solid phase. Typical precursors are metal alkoxides and metal chlorides, which undergo hydrolysis and polycondensation reactions to form a colloid. The basic structure or morphology of the solid phase can range anywhere from discrete colloidal particles to continuous chain-like polymer networks.

The term *colloid* is used primarily to describe a broad range of solid-liquid (and/or liquid-liquid) mixtures, all of which contain distinct solid (and/or liquid) particles which are dispersed to various degrees in a liquid medium. The term is specific to the size of the individual particles, which are larger than atomic dimensions but small enough to exhibit Brownian motion. If the particles are large enough, then their dynamic behavior in any given period of time in suspension would be governed by forces of gravity and sedimentation. But if they are small enough to be colloids, then their irregular motion in suspension can be attributed to the collective bombardment of a myriad of thermally agitated molecules in the liquid suspending medium, as described originally by Albert Einstein in his dissertation. Einstein concluded that this erratic behavior could adequately be described using the theory of Brownian motion, with sedimentation being a possible long-term result. This critical size range (or particle diameter) typically ranges from tens of angstroms (10^{-10} m) to a few micrometres (10^{-6} m).

- Under certain chemical conditions (typically in base-catalyzed sols), the particles may grow to sufficient size to become colloids, which are affected both by sedimentation and forces of gravity. Stabilized suspensions of such sub-micrometre spherical particles may eventually result in their self-assembly—yielding highly ordered microstructures reminiscent of the prototype colloidal crystal: precious opal.
- Under certain chemical conditions (typically in acid-catalyzed sols), the interparticle forces have sufficient strength to cause considerable aggregation and/or flocculation prior to their growth. The formation of a more open continuous network of low density polymers exhibits certain advantages with regard to physical properties in the formation of high performance glass and glass/ceramic components in 2 and 3 dimensions.

In either case (discrete particles or continuous polymer network) the sol evolves then towards the formation of an inorganic network containing a liquid phase (gel). Formation of a metal oxide involves connecting the metal centers with oxo (M-O-M) or hydroxo (M-OH-M) bridges, therefore generating metal-oxo or metal-hydroxo polymers in solution.

In both cases (discrete particles or continuous polymer network), the drying process serves to remove the liquid phase from the gel, yielding a micro-porous amorphous glass or micro-crystalline ceramic. Subsequent thermal treatment (firing) may be performed in order to favor further polycondensation and enhance mechanical properties.

With the viscosity of a sol adjusted into a proper range, both optical quality glass fiber and refractory ceramic fiber can be drawn which are used for fiber optic sensors and thermal insulation, respectively. In addition, uniform ceramic powders of a wide range of chemical composition can be formed by precipitation.

Polymerization

The Stöber process is a well-studied example of polymerization of an alkoxide, specifically TEOS. The chemical formula for TEOS is given by $Si(OC_2H_5)_4$, or $Si(OR)_4$, where the alkyl group $R = C_2H_5$. Alkoxides are ideal chemical precursors for sol–gel synthesis because they react readily with water. The reaction is called hydrolysis, because a hydroxyl ion becomes attached to the silicon atom as follows:

$$Si(OR)_4 + H_2O \rightarrow HO\text{-}Si(OR)_3 + R\text{-}OH$$

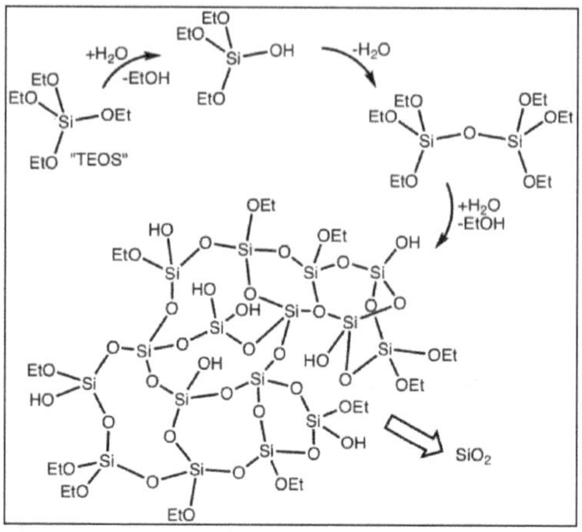

Simplified representation of the condensation induced by hydrolysis of TEOS.

Depending on the amount of water and catalyst present, hydrolysis may proceed to completion to silica:

$$Si(OR)_4 + 2\,H_2O \rightarrow SiO_2 + 4\,R\text{-}OH$$

Complete hydrolysis often requires an excess of water and/or the use of a hydrolysis catalyst such as acetic acid or hydrochloric acid. Intermediate species including $[(OR)_2\text{–}Si\text{–}(OH)_2]$ or $[(OR)_3\text{–}Si\text{–}(OH)]$ may result as products of partial hydrolysis reactions. Early intermediates result from two partially hydrolyzed monomers linked with a siloxane [Si–O–Si] bond:

$$(OR)_3\text{-}Si\text{-}OH + HO\text{-}Si\text{-}(OR)_3 \rightarrow [(OR)_3\,Si\text{-}O\text{-}Si(OR)_3] + H\text{-}O\text{-}H$$

or

$$(OR)_3\text{-}Si\text{-}OR + HO\text{-}Si\text{-}(OR)_3 \rightarrow [(OR)_3\,Si\text{-}O\text{-}Si(OR)_3] + R\text{-}OH$$

Thus, polymerization is associated with the formation of a 1-, 2-, or 3-dimensional network of siloxane [Si–O–Si] bonds accompanied by the production of H–O–H and R–O–H species.

By definition, condensation liberates a small molecule, such as water or alcohol. This type of reaction can continue to build larger and larger silicon-containing molecules by the process of polymerization. Thus, a polymer is a huge molecule (or macromolecule) formed from hundreds or thousands of units called monomers. The number of bonds that a monomer can form is called its functionality. Polymerization of silicon alkoxide, for instance, can lead to complex branching of the polymer, because a fully hydrolyzed monomer $Si(OH)_4$ is tetrafunctional (can branch or bond in 4 different directions). Alternatively, under certain conditions (e.g., low water concentration) fewer than 4 of the OR or OH groups (ligands) will be capable of condensation, so relatively little branching will occur. The mechanisms of hydrolysis and condensation, and the factors that bias the structure toward linear or branched structures are the most critical issues of sol–gel science and technology. This reaction is favored in both basic and acidic conditions.

Sono-Ormosil

Sonication is an efficient tool for the synthesis of polymers. The cavitational shear forces, which stretch out and break the chain in a non-random process, result in a lowering of the molecular weight and poly-dispersity. Furthermore, multi-phase systems are very efficient dispersed and emulsified, so that very fine mixtures are provided. This means that ultrasound increases the rate of polymerisation over conventional stirring and results in higher molecular weights with lower polydispersities. Ormosils (organically modified silicate) are obtained when silane is added to gel-derived silica during sol–gel process. The product is a molecular-scale composite with improved mechanical properties. Sono-Ormosils are characterized by a higher density than classic gels as well as an improved thermal stability. An explanation therefore might be the increased degree of polymerization.

Pechini Process

For single cation systems like SiO_2 and TiO_2, hydrolysis and condensation processes naturally give rise to homogenous compositions. For systems involving multiple cations, such as strontium titanate, $SrTiO_3$ and other perovskite systems, the concept of steric immobilisation becomes relevant. To avoid the formation of multiple phases of binary oxides as the result of differing hydrolysis and condensation rates, the entrapment of cations in a polymer network is an effective approach, generally termed the Pechini Process. In this process, a chelating agent is used, most often citric acid, to surround aqueous cations and sterically entrap them. Subsequently a polymer network is formed to immobilize the chelated cations in a gel or resin. This is most often achieved by poly-esterification using ethylene glycol. The resulting polymer is then combusted under oxidising conditions to remove organic content and yield a product oxide with homogeneously dispersed cations.

Nanomaterials

In the processing of fine ceramics, the irregular particle sizes and shapes in a typical powder often lead to non-uniform packing morphologies that result in packing density variations in the powder compact. Uncontrolled flocculation of powders due to attractive van der Waals forces can also give rise to microstructural heterogeneities.

Differential stresses that develop as a result of non-uniform drying shrinkage are directly related to the rate at which the solvent can be removed, and thus highly dependent upon the distribution of porosity. Such stresses have been associated with a plastic-to-brittle transition in consolidated bodies, and can yield to crack propagation in the unfired body if not relieved.

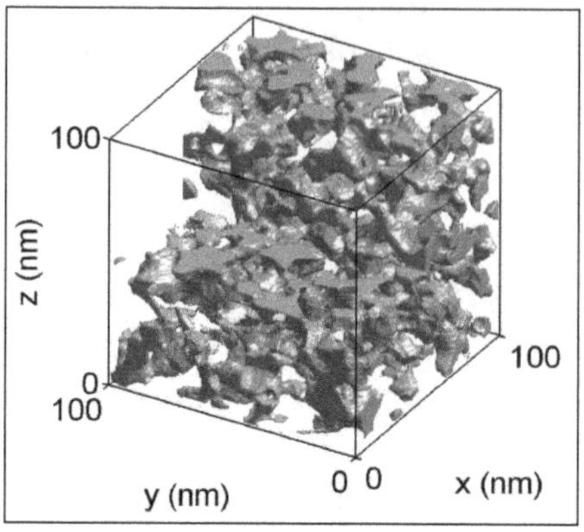

Nanostructure of a resorcinol-formaldehyde gel reconstructed from small-angle X-ray scattering. This type of disordered morphology is typical of many sol–gel materials.

In addition, any fluctuations in packing density in the compact as it is prepared for the kiln are often amplified during the sintering process, yielding heterogeneous densification. Some pores and other structural defects associated with density variations have been shown to play a detrimental role in the sintering process by growing and thus limiting end-point densities. Differential stresses arising from heterogeneous densification have also been shown to result in the propagation of internal cracks, thus becoming the strength-controlling flaws.

It would therefore appear desirable to process a material in such a way that it is physically uniform with regard to the distribution of components and porosity, rather than using particle size distributions which will maximize the green density. The containment of a uniformly dispersed assembly of strongly interacting particles in suspension requires total control over particle-particle interactions. Monodisperse colloids provide this potential.

Monodisperse powders of colloidal silica, for example, may therefore be stabilized sufficiently to ensure a high degree of order in the colloidal crystal or polycrystalline colloidal solid which results from aggregation. The degree of order appears to be limited by the time and space allowed for longer-range correlations to be established. Such defective polycrystalline structures would appear to be the basic elements of nanoscale materials science, and, therefore, provide the first step in developing a more rigorous understanding of the mechanisms involved in microstructural evolution in inorganic systems such as sintered ceramic nanomaterials.

Applications

The applications for sol gel-derived products are numerous. For example, scientists have used it to produce the world's lightest materials and also some of its toughest ceramics.

Protective Coatings

One of the largest application areas is thin films, which can be produced on a piece of substrate by spin coating or dip-coating. Protective and decorative coatings, and electro-optic components can be applied to glass, metal and other types of substrates with these methods. Cast into a mold, and with further drying and heat-treatment, dense ceramic or glass articles with novel properties can be formed that cannot be created by any other method. Other coating methods include spraying, electrophoresis, inkjet printing, or roll coating.

Thin Films and Fibers

With the viscosity of a sol adjusted into a proper range, both optical and refractory ceramic fibers can be drawn which are used for fiber optic sensors and thermal insulation, respectively. Thus, many ceramic materials, both glassy and crystalline, have found use in various forms from bulk solid-state components to high surface area forms such as thin films, coatings and fibers.

Nanoscale Powders

Ultra-fine and uniform ceramic powders can be formed by precipitation. These powders of single and multiple component compositions can be produced on a nanoscale particle size for dental and biomedical applications. Composite powders have been patented for use as agrochemicals and herbicides. Powder abrasives, used in a variety of finishing operations, are made using a sol–gel type process. One of the more important applications of sol–gel processing is to carry out zeolite synthesis. Other elements (metals, metal oxides) can be easily incorporated into the final product and the silicate sol formed by this method is very stable.

Another application in research is to entrap biomolecules for sensory (biosensors) or catalytic purposes, by physically or chemically preventing them from leaching out and, in the case of protein or chemically-linked small molecules, by shielding them from the external environment yet allowing small molecules to be monitored. The major disadvantages are that the change in local environment may alter the functionality of the protein or small molecule entrapped and that the synthesis step may damage the protein. To circumvent this, various strategies have been explored, such as monomers with protein friendly leaving groups (e.g. glycerol) and the inclusion of polymers which stabilize protein (e.g. PEG).

Other products fabricated with this process include various ceramic membranes for microfiltration, ultrafiltration, nanofiltration, pervaporation, and reverse osmosis. If the liquid in a wet gel is removed under a supercritical condition, a highly porous and extremely low density material called aerogel is obtained. Drying the gel by means of low temperature treatments (25-100 °C), it is possible to obtain porous solid matrices called xerogels. In addition, a sol–gel process was developed in the 1950s for the production of radioactive powders of UO_2 and ThO_2 for nuclear fuels, without generation of large quantities of dust.

Opto-mechanical

Macroscopic optical elements and active optical components as well as large area hot mirrors, cold mirrors, lenses, and beam splitters all with optimal geometry can be made quickly and at low cost via the sol–gel route. In the processing of high performance ceramic nanomaterials with superior opto-mechanical properties under adverse conditions, the size of the crystalline grains is determined largely by the size of the crystalline particles present in the raw material during the synthesis or formation of the object. Thus a reduction of the original particle size well below the wavelength of visible light (~500 nm) eliminates much of the light scattering, resulting in a translucent or even transparent material.

Furthermore, results indicate that microscopic pores in sintered ceramic nanomaterials, mainly trapped at the junctions of microcrystalline grains, cause light to scatter and prevented true transparency. It has been observed that the total volume fraction of these nanoscale pores (both intergranular and intragranular porosity) must be less than 1% for high-quality optical transmission. I.E. The density has to be 99.99% of the theoretical crystalline density.

Medicine

Unique properties of the sol–gel provide the possibility of their use for a variety of medical applications. A sol–gel processed alumina can be used as a carrier for the sustained delivery of drugs and as an established wound healer. A marked decrease in scar size was observed because of the wound healing composite including sol–gel processed alumina. A novel approach to thrombolysis treatment is possible by developing a new family of injectable composites: plasminogen activator entrapped within alumina.

Chemical Transport Reaction

Gold crystals grown by chemical transport using chlorine as the transport agent.

In chemistry, a chemical transport reaction describes a process for purification and crystallization of non-volatile solids. The process is also responsible for certain aspects of mineral growth from the effluent of volcanoes. The technique is distinct from chemical vapor deposition, which usually entails decomposition of molecular precursors (e.g. $SiH_4 \rightarrow Si + 2H_2$) and which gives conformal coatings. The technique, which was popularized by Harald Schäfer, entails the reversible conversion of nonvolatile elements and chemical compounds into volatile derivatives. The volatile derivative migrates throughout a sealed reactor, typically a sealed and evacuated glass tube heated in a tube furnace. Because the tube is under a temperature gradient, the volatile derivative reverts to the parent solid and the transport agent is released at the end opposite to which it originated. The transport agent is thus catalytic. The technique requires that the two ends of the tube (which contains the sample to be crystallized) be maintained at different temperatures. So-called two-zone tube furnaces are employed for this purpose. The method derives from the Van Arkel de Boer process which was used for the purification of titanium and vanadium and uses iodine as the transport agent.

Crystals of titanium grown using the Van Arkel-de Boer process with I_2 as the transport agent.

Cases of the Exothermic and Endothermic Reactions of the Transporting Agent

Transport reactions are classified according to the thermodynamics of the reaction between the solid and the transporting agent. When the reaction is exothermic, then the solid of interest is transported from the cooler end (which can be quite hot) of the reactor to a hot end, where the equilibrium constant is less favorable and the crystals grow. The reaction of molybdenum dioxide with the transporting agent iodine is an exothermic process, thus the MoO_2 migrates from the cooler end (700 °C) to the hotter end (900 °C):

$$MoO_2 + I_2 \rightleftharpoons MoO_2I_2 \; \Delta H_{rxn} < 0 \left(\text{exothermic}\right)$$

Using 10 milligrams of iodine for 4 grams of the solid, the process requires several days.

Alternatively, when the reaction of the solid and the transport agent is endothermic, the solid is transported from a hot zone to a cooler one. For example:

$$Fe_2O_3 + 6\,HCl \rightleftharpoons Fe_2Cl_6 + 3\,H_2O \; \Delta H_{rxn} > 0 \left(\text{endothermic}\right)$$

The sample of iron(III) oxide is maintained at 1000 °C, and the product is grown at 750 °C. HCl is the transport agent. Crystals of hematite are reportedly observed at the mouths of volcanoes because of chemical transport reactions whereby volcanic hydrogen chloride volatilizes iron(III) oxides.

Halogen Lamp

A similar reaction like that of MoO_2 is used in halogen lamps. The tungsten is evaporated from the tungsten filament and converted with traces of oxygen and iodine into the WO_2I_2, at the high temperatures near the filament the compound decomposes back to tungsten, oxygen and iodine.

$$WO_2 + I_2 \rightleftharpoons WO_2I_2, \Delta H_{rxn} < 0 \text{ (exothermic)}$$

Solid-gas Reaction

Solid-gas reaction is a type of solid state reaction that is performed by exposing a reactive solid to a stream of reactive gas. Depending on reagent reactivity, solid-gas reactions can be performed at reduced, elevated or at room temperature. As well as solid state reactions, solid-gas reactions offer reduced costs, decreased amounts of chemical waste, time savings and, sometimes, perceptively hygher yields.

References

- Solid-State-Reaction: encyclopedia2.thefreedictionary.com, Retrieved 02 February, 2020
- Reaction-mechanisms, boundless-chemistry: courses.lumenlearning.com, Retrieved 23 March, 2020
- Dhara, S. (2007). "Formation, Dynamics, and Characterization of Nanostructures by Ion Beam Irradiation". Critical Reviews in Solid State and Materials Sciences. 32 (1): 1–50. Bibcode:2007CRSSM..32....1D. doi:10.1080/10408430601187624
- Brinker, C. J.; G. W. Scherer (1990). Sol-Gel Science: The Physics and Chemistry of Sol-Gel Processing. Academic Press. ISBN 978-0-12-134970-7
- Michael Binnewies, Robert Glaum, Marcus Schmidt, Peer Schmidt "Chemical Vapor Transport Reactions – A Historical Review" Zeitschrift für anorganische und allgemeine Chemie 2013, Volume 639, pages 219–229. doi:10.1002/zaac.201300048
- Solid-gas-reaction, applications: labochema.com, Retrieved 14 April, 2020

5
Crystal Structure and Defects

Crystal defects refer to the imperfections in the regular patterns of the atoms in crystalline solids. Different defects include vacancy defect, Schottky defect, Frenkel defect, interstitial defect, Stone-Wales defect, etc. This chapter has been carefully written to provide an easy understanding of the subject of crystal defects and non-stoichiometry.

Crystal Structure

Crystal Structure is obtained by attaching atoms, groups of atoms or molecules. This structure occurs from the intrinsic nature of the constituent particles to produce symmetric patterns. A small group of a repeating pattern of the atomic structure is known as the unit cell of the structure. A unit cell is the building block of the crystal structure and it also explains in detail the entire crystal structure and symmetry with the atom positions along with its principal axes. The length, edges of principal axes and the angle between the unit cells are called lattice constants or lattice parameters.

Unit Cells

The structure of solids can be described as if they were three-dimensional analogs of a piece of wallpaper. Wallpaper has a regular repeating design that extends from one edge to the other. Crystals have a similar repeating design, but in this case the design extends in three dimensions from one edge of the solid to the other.

The simplest repeating unit in a crystal is called a unit cell. Each unit cell is defined in terms of lattice points-the points in space about which the particles are free to vibrate in a crystal.

The structures of the unit cell for a variety of salts are shown below.

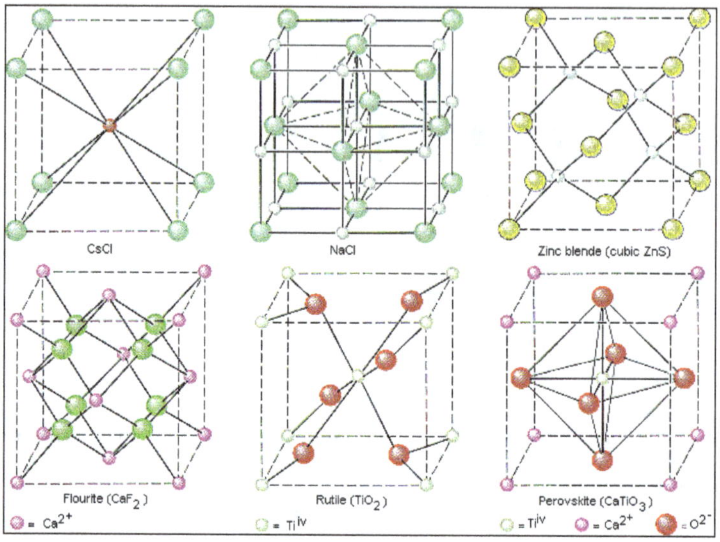

In 1850, Auguste Bravais showed that crystals could be divided into 14 unit cells, which meet the following criteria:

- The unit cell is the simplest repeating unit in the crystal.
- Opposite faces of a unit cell are parallel.
- The edge of the unit cell connects equivalent points.

The 14 Bravais unit cells are shown in the figure below.

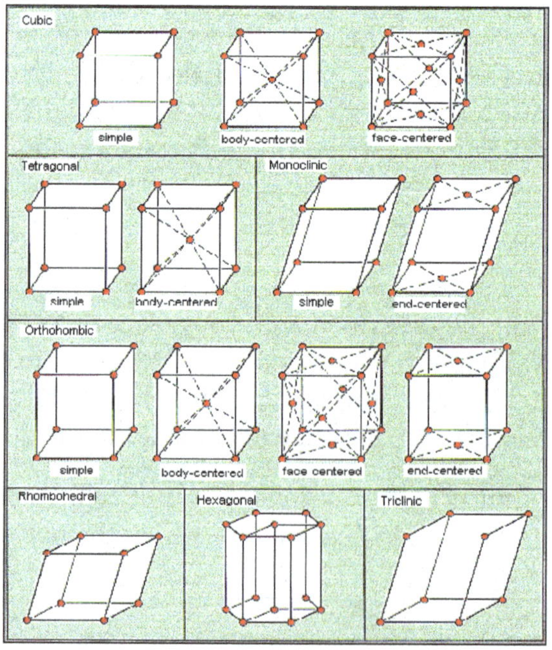

These unit cells fall into seven categories, which differ in the three unit-cell edge lengths (a, b, and c) and three internal angles (a, � and g).

Table: The seven categories of Bravais unit cells.

Category	Edge Lengths	Internal Angles
Cubic	(a = b = c)	(a = x/i> = g = 90°)
Tetragonal	(a = b ≠ c)	(a = x/i> = g = 90°)
Monoclinic	(a ≠ b ≠ c)	(a = x/i> = 90° ≠ g)
Orthorhombic	(a ≠ b ≠ c)	(a = x/i> = g = 90°)
Rhombohedral	(a = b = c)	(a = x/i> = g ≠ 90°)
Hexagonal	(a = b ≠ c)	(a = x/i> = 90°, g = 120°)
Triclinic	(a ≠ b ≠ c)	(a ≠ x/i> ≠ g ≠ 90°)

We will focus on the cubic category, which includes the three types of unit cells--simple cubic, body-centered cubic, and face-centered cubic.

These unit cells are important for two reasons. First, a number of metals, ionic solids, and intermetallic compounds crystallize in cubic unit cells. Second, it is relatively easy to do calculations with these unit cells because the cell-edge lengths are all the same and the cell angles are all 90.

The simple cubic unit cell is the simplest repeating unit in a simple cubic structure. Each corner of the unit cell is defined by a lattice point at which an atom, ion, or molecule can be found in the crystal. By convention, the edge of a unit cell always connects equivalent points. Each of the eight corners of the unit cell therefore must contain an identical particle. Other particles can be present on the edges or faces of the unit cell, or within the body of the unit cell. But the minimum that must be present for the unit cell to be classified as simple cubic is eight equivalent particles on the eight corners.

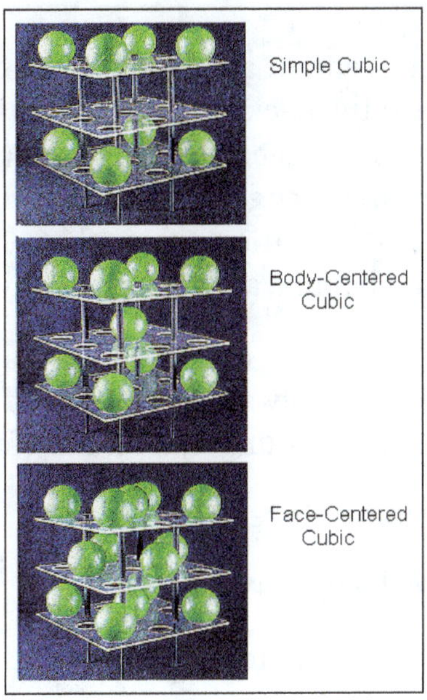

The body-centered cubic unit cell is the simplest repeating unit in a body-centered cubic structure. Once again, there are eight identical particles on the eight corners of the unit cell. However, this time there is a ninth identical particle in the center of the body of the unit cell.

The face-centered cubic unit cell also starts with identical particles on the eight corners of the cube. But this structure also contains the same particles in the centers of the six faces of the unit cell, for a total of 14 identical lattice points.

The face-centered cubic unit cell is the simplest repeating unit in a cubic closest-packed structure. In fact, the presence of face-centered cubic unit cells in this structure explains why the structure is known as cubic closest-packed.

The lattice points in a cubic unit cell can be described in terms of a three-dimensional graph. Because all three cell-edge lengths are the same in a cubic unit cell, it doesn't matter what orientation is used for the a, b, and c axes. For the sake of argument, we'll define the a axis as the vertical axis of our coordinate system.

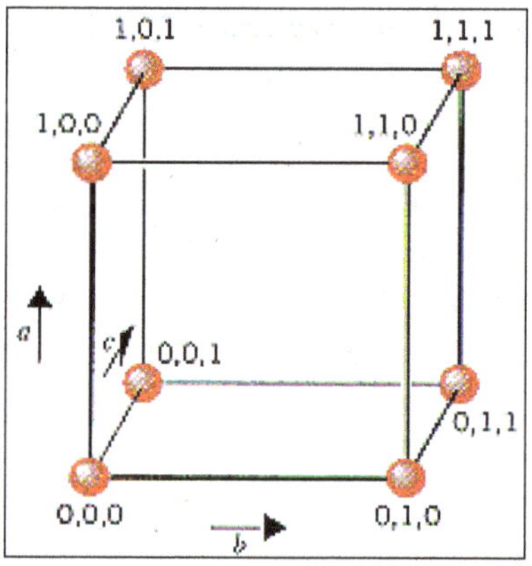

The b axis will then describe movement across the front of the unit cell, and the c axis will represent movement toward the back of the unit cell. Furthermore, we'll arbitrarily define the bottom left corner of the unit cell as the origin (0,0,0). The coordinates 1,0,0 indicate a lattice point that is one cell-edge length away from the origin along the a axis. Similarly, 0,1,0 and 0,0,1 represent lattice points that are displaced by one cell-edge length from the origin along the b and c axes, respectively.

Thinking about the unit cell as a three-dimensional graph allows us to describe the structure of a crystal with a remarkably small amount of information. We can specify the structure of cesium chloride, for example, with only four pieces of information:

- CsCl crystallizes in a cubic unit cell.
- The length of the unit cell edge is 0.4123 nm.
- There is a Cl^- ion at the coordinates 0,0,0.

- There is a Cs^+ ion at the coordinates 1/2,1/2,1/2.

Because the cell edge must connect equivalent lattice points, the presence of a Cl^- ion at one corner of the unit cell (0,0,0) implies the presence of a Cl^- ion at every corner of the cell. The coordinates 1/2,1/2,1/2 describe a lattice point at the center of the cell. Because there is no other point in the unit cell that is one cell-edge length away from these coordinates, this is the only Cs^+ ion in the cell. CsCl is therefore a simple cubic unit cell of Cl^- ions with a Cs^+ in the center of the body of the cell.

Bragg's Law

Experimental Method of Determining Crystal Structure

X-ray diffraction and Bragg's Equation: Crystal structure has been obtained by studying on the diffraction of X-rays by solids. A crystal, having constituents particles arranged in planes at very small distances in three dimension array, acts as diffraction grating for X– rays which have wavelengths of the same order as the spacing in crystal.

When a beam of X-rays passes through a crystalline solid, each atom in the beam scatters some of the radiations. If waves are on same phase means if their peak and trough coincides they add together to give a wave of greater amplitude. This enhancement of intensity is called constructive interference. If waves are out of phase, they cancel.

This cancellation is called destructive interference.

Thus X– ray diffraction results from the scattering of X-rays by a regular arrangement of atoms or ions.

Bragg's equation: Study of internal structure of crystal can be done with the help of X-rays. The distance of the constituent particles can be determined from diffraction value by Bragg's equation, $n\lambda = 2d \sin\theta$ where, l = Wave length of X-rays, n = order of diffraction, θ = Angle of reflection, d = Distance between two parallel surfaces

The above equation is known as Bragg's equation or Bragg's law. The reaction corresponding to n = 1 (for a given family of planes) is called first order refection; the reflection corresponding to n = 2 is the second order reflection and so on. Thus by measuring n (the order of reflection of the X-rays) and the incidence angle q, we can know d/λ.

$$\frac{d}{\lambda} = \frac{n}{2\sin\theta}$$

From this, d can be calculated if l is known and vice versa. In X-ray reactions, n is generally set as equal to 1. Thus,

Bragg's equation may alternatively be written as:

$$\lambda = 2d\sin\theta = 2d_{hkl}\sin\theta$$

Where d_{hkl} denotes the perpendicular distance between adjacent planes with the indices hkl.

Crystal Systems

A Crystal System refers to one of the many classes of crystals, space groups, and lattices. In crystallography terms, lattice system and crystal, the system are associated with each other with a slight difference. Based on their point groups crystals and space groups are divided into seven crystal systems.

The Seven Crystal Systems is an approach for classification depending upon their lattice and atomic structure. The atomic lattice is a series of atoms that are organized in a symmetrical pattern. With the help of the lattice, it is possible to determine the appearance and physical properties of the stone. It is possible to identify so as to which crystal system they belong to. In a Cubic System crystals are said to represent the element earth. They are Seven Crystal Systems and are stated below with illustrated examples.

The Seven Crystal Systems

Triclinic System

All three axes are inclined towards each other, and they are of the same length. Based on the three inclined angles the various forms of crystals are in the paired faces. Some of the standard Triclinic Systems include Labradorite, Amazonite, Kyanite, Rhodonite, Aventurine Feldspar, and Turquoise.

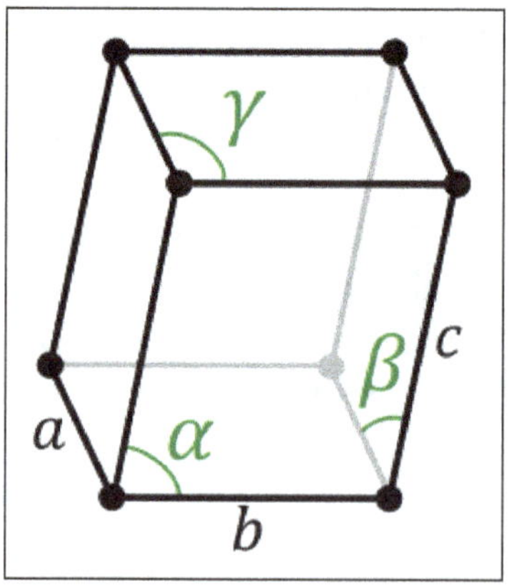

Triclinic System.

Monoclinic System

It comprises of three axes where two are at right angles to each other, and the third axis is inclined. All three axes are of different length. Based on the inner structure monoclinic system includes Basal pinacoids and prisms with inclined end faces. Some of the examples include Diopside, Petalite, Kunzite, Gypsum, Hiddenite, Howlite, Vivianite and more.

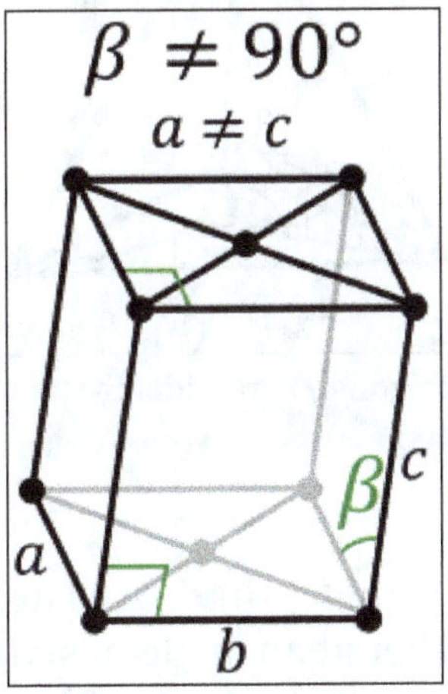

Monoclinic System.

Orthorhombic System

It comprises of three axes and is at right angles to each other. There are different lengths. Based on their Rhombic structure orthorhombic system includes various crystal shapes namely pyramids, double pyramids, rhombic pyramids, and pinacoids. Some of the common orthorhombic crystals include Topaz, Tanzanite, Iolite, Zoisite, Danburite and more.

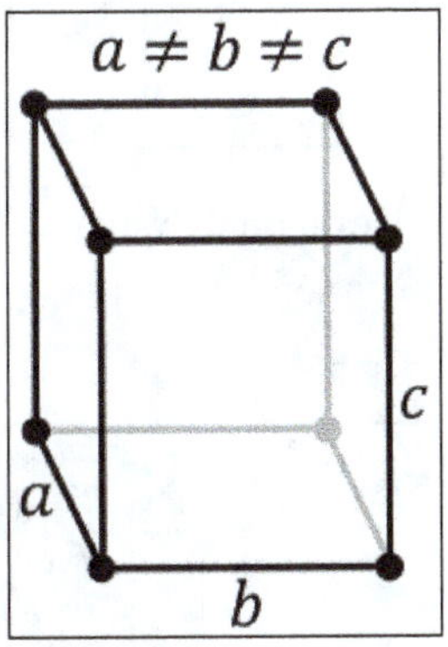

Orthorhombic System.

Trigonal System

Angles and axis in a trigonal system are similar to Hexagonal Systems. At the base of a hexagonal system (cross-section of a prism), there will be six sides. In the trigonal system (base cross-section) there will be three sides. Crystal shapes in a trigonal system include three-sided pyramids, Scalenohedral and Rhombohedra. Some of the typical examples include Ruby, Quartz, Calcite, Agate, Jasper, Tiger's Eyes and more.

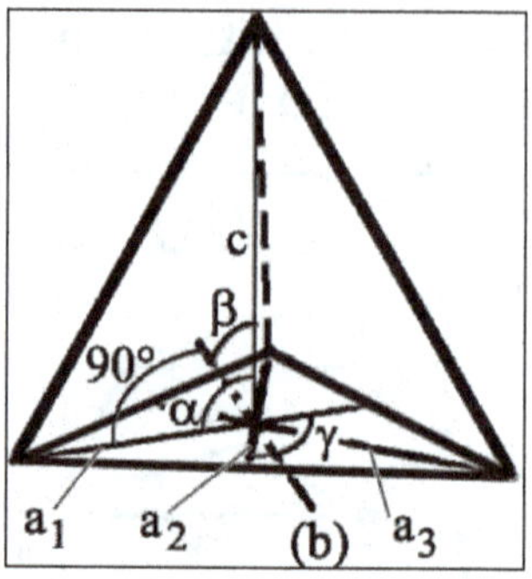

Trigonal/Pyramidal System.

Hexagonal System

It comprises of four axes. Among them, three axes are of the same length and are on one plane. They intersect each other at an angle of sixty degrees. The fourth axis intersects other axes at right angles. Crystal shapes of hexagonal systems include Double Pyramids, Double-Sided Pyramids, and Four-Sided Pyramids. Example: Beryl, Cancrinite, Apatite, Sugilite, etc.

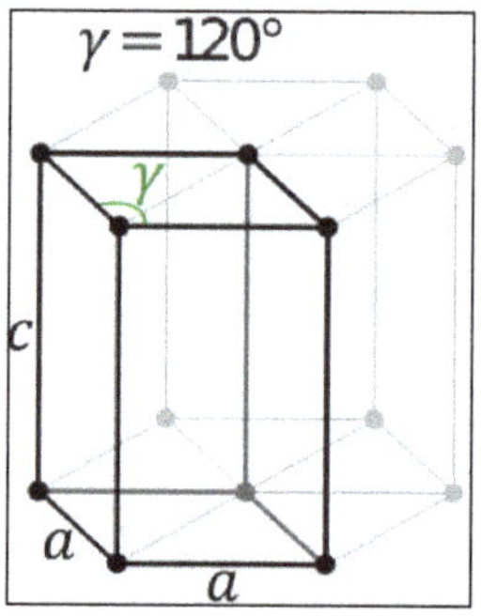

Hexagonal System.

Tetragonal Systems

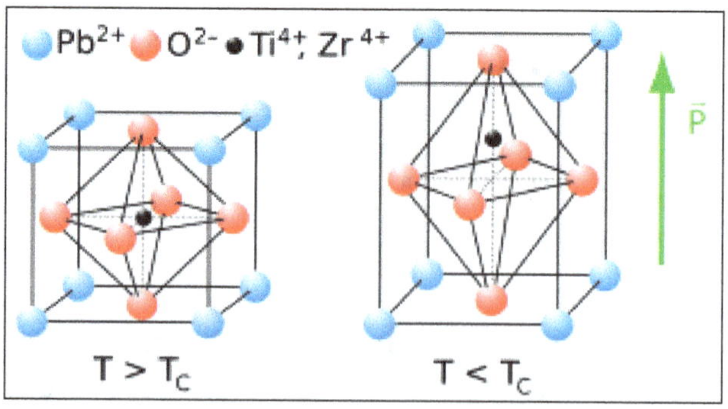

Perovskite – Tetragonal Systems.

It consists of three axes. The main axis varies in length; it can either be short or long. The two-axis lie in the same plane and are of the same length. Based on the rectangular inner structure the shapes of crystal in tetragonal include double and eight-sided pyramids, four-sided prism, trapezohedrons, and pyrite.

Cubic System

All three angles intersect at right angles and are of equal length. Crystal shapes of a cubic system based on inner structure (square) include octahedron, cube, and Hexaciscoherdron. Example: Silver, Garnet, Gold, and Diamond.

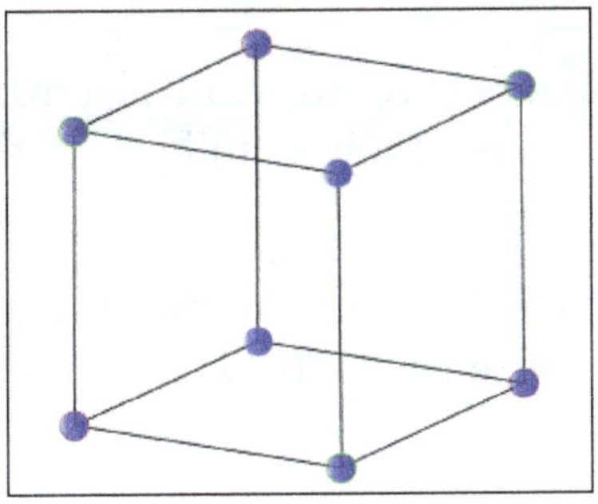

Cubic System.

Atomic Structure

Atomic structure refers to the structure of atom comprising of a nucleus (center) in which the protons (positively charged) and neutrons (neutral) are present. The negatively charged particles called electrons revolve around the center of the nucleus.

The history of atomic structure and quantum mechanics dates back to the times of Democritus, the man who first proposed that matter is composed of atoms.

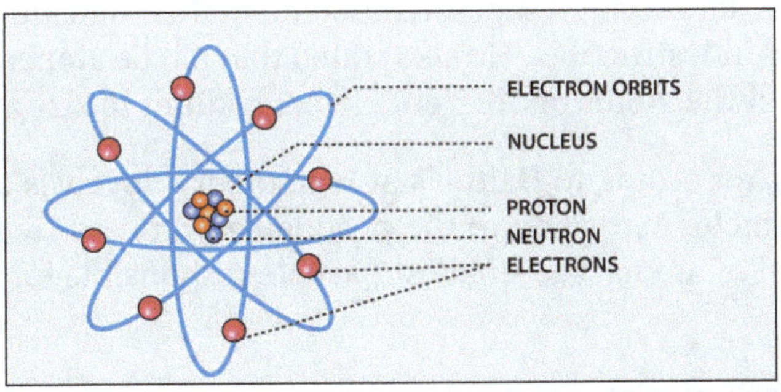

The Structure of Atom – Electrons, Protons, and Neutrons.

The study about the structure of atom gives a great insight into the entire class of chemical reactions, bonds, and their physical properties. The first scientific theory of atomic structure was proposed by John Dalton in 1800s.

The advances in atomic structure and quantum mechanics have led to the discovery of other fundamental particles. The discovery of subatomic particles has been the base for many other discoveries and inventions.

Atomic Structure

The atomic structure of an element refers to the constitution of its nucleus and the arrangement of the electrons around it. Primarily, the atomic structure of matter is made up of protons, electrons, and neutrons.

The protons and neutrons make up the nucleus of the atom, which is surrounded by the electrons belonging to the atom. The atomic number of an element describes the total number of protons in its nucleus.

Neutral atoms have equal numbers of protons and electrons. However, atoms may gain or lose electrons in order to increase their stability, and the resulting charged entity is called an ion.

Atoms of different elements have different atomic structures because they contain different numbers of protons and electrons. This is the reason for the unique characteristics of different elements.

Atomic Models

In the 18th and 19th centuries, many scientists attempted to explain the structure of the atom with the help of atomic models. Each of these models had their own merits and demerits and were pivotal to the development of the modern atomic model. The most notable contributions to the field were by the scientists John Dalton, J.J. Thomson, Ernest Rutherford, and Niels Bohr.

Dalton's Atomic Theory

The English chemist John Dalton suggested that all matter is made up of atoms, which were indivisible and indestructible. He also stated that all the atoms of an element were exactly the same, but the atoms of different elements differ in size and mass.

Chemical reactions, according to Dalton's atomic theory, involves a rearrangement of atoms to form products. According to the postulates proposed by Dalton, the atomic structure comprised of atoms, the smallest particle responsible for the chemical reactions to occur.

The following are the postulates of his theory:

- Every matter is made up of atoms.
- Atoms are indivisible.
- Specific elements have only one type of atoms in them.
- Each atom has its own constant mass that varies from element to element.
- Atoms undergo rearrangement during a chemical reaction.

- Atoms can neither be created nor be destroyed but can be transformed from one form to another.

Dalton's atomic theory successfully explained the Laws of chemical reactions, namely, the Law of conservation of mass, Law of constant properties, Law of multiple proportions, and Law of reciprocal proportions.

Demerits of Dalton's Atomic Theory

- The theory was unable to explain the existence of isotopes.
- Nothing about the structure of atom was appropriately is explained.
- Later, the scientists discovered particles inside the atom that proved, the atoms are divisible.

The discovery of particles inside atoms led to a better understanding of chemical species, these particles inside the atoms are called subatomic particles.

The discovery of various subatomic particles is as follows.

Thomson Atomic Model

The English chemist Sir Joseph John Thomson put forth his model describing the atomic structure in the early 1900s.

He was later awarded the Nobel prize for the discovery of "electrons". His work is based on an experiment called cathode ray experiment. The construction of working of the experiment is as follows:

Cathode Ray Experiment

It has a tube made of glass which has two openings, one for the vacuum pump and the other for the inlet through which a gas is pumped in.

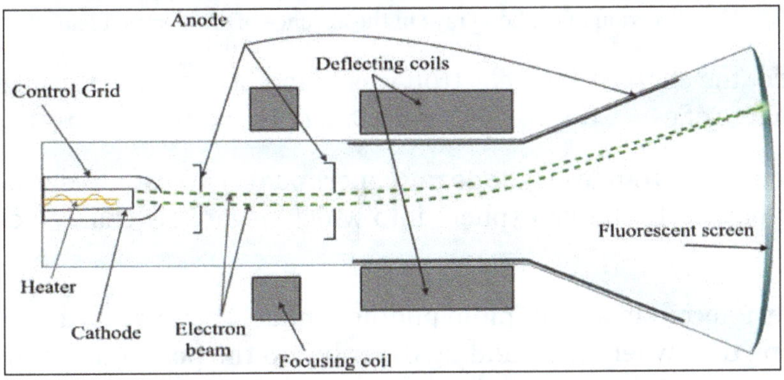

Cathode Ray Experiment: The Discovery of Electrons.

The role of the vacuum pump is to maintain "partial vacuum" inside the glass chamber. A high voltage power supply is connected using electrodes i.e. cathode and Anode is fitted inside the glass tube.

Observations of Cathode Ray Experiment

- When a high voltage power supply is switched on, there were rays emerging from the cathode towards the anode. This was confirmed by the 'Fluorescent spots' on the ZnS screen used. These rays were called "Cathode Rays".
- When an external electric field is applied, the cathode rays get deflected towards the positive electrode, but in the absence of electric field, they travel in a straight line.
- When rotor Blades are placed in the path of the cathode rays, they seem to rotate. This proves that the cathode rays are made up of particles of a certain mass so that they have some energy.
- With all this evidence, Thompson concluded that cathode rays are made of negatively charged particle called "electrons".
- On applying the electric and magnetic field upon the cathode rays (electrons), Thomson found the charge to mass ratio (e/m) of electrons. (e/m) for electron: 17588×1011 e/bg.

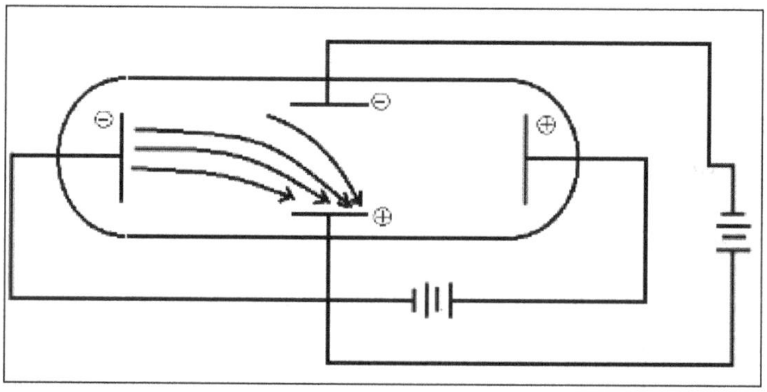

The direction of Cathode rays in the absence of an Electric Field.

From this ratio, the charge of the electron was found by Mullikin through oil drop experiment. [Charge of e- = 1.6×10-16 C and Mass of e- = 9.1093×10-31 kg].

Based on conclusions from his cathode ray experiment, Thomson described the atomic structure as a positively charged sphere into which negatively charged electrons were embedded.

It is commonly referred to as the "plum pudding model" because it can be visualized as a plum pudding dish where the pudding describes to the positively charged atom and the plum pieces describe the electrons.

Thomson's atomic structure described atoms as electrically neutral, i.e. the positive and the negative charges were of equal magnitude.

Limitations of Thomson's Atomic Structure: Thomson's atomic model does not clearly explain the stability of an atom. Also, further discoveries of other subatomic particles, couldn't be placed inside his atomic model.

Rutherford Atomic Theory

Rutherford, a student of J. J. Thomson modified the atomic structure with the discovery of another subatomic particle called "Nucleus". His atomic model is based on the Alpha ray scattering experiment.

Alpha Ray Scattering Experiment

- Construction:
  - A very thin gold foil of 1000 atoms thick is taken.
  - Alpha rays (doubly charged Helium He2+) were made to bombard the gold foil.
  - ZnS screen is placed behind the gold foil.
- Observations:
  - Most of the rays just went through the gold foil making scintillations (bright spots) in the ZnS screen.
  - A few rays got reflected after hitting the gold foil.
  - One in 1000 rays got reflected by an angle of 180° (retraced path) after hitting the gold foil.
- Conclusions:
  - Since most rays passed through, Rutherford concluded that most of the space inside the atom is empty.
  - Few rays got reflected because of the repulsion of its positive with some other positive charge inside the atom.
  - 1/1000th of rays got strongly deflected because of a very strong positive charge in the center of the atom he called this strong positive charge as "nucleus".
  - He said most of the charge and mass of the atom resides in the Nucleus.

Rutherford's Structure of Atom

Based on the above observations and conclusions, Rutherford proposed his own atomic structure which is as follows:

- The nucleus is at the center of an atom, where most of the charge and mass are concentrated.
- Atomic structure is spherical.
- Electrons revolve around the nucleus in a circular orbit, similar to the way planets orbit the sun.

Limitations of Rutherford Atomic Model

- If electrons have to revolve around the nucleus, they will spend energy and that too against the strong force of attraction from the nucleus, a lot of energy will be spent by the electrons and eventually, they will lose all their energy and will fall into the nucleus so the stability of atom is not explained.
- If electrons continuously revolve around the 'nucleus, the type of spectrum expected is a continuous spectrum. But in reality, what we see is a line spectrum.

Subatomic Particles

Protons

- Protons are positively charged subatomic particles. The charge of a proton is 1e, which corresponds to approximately 1.602×10^{-19}.
- The mass of a proton is approximately 1.672×10^{-24}.
- Protons are over 1800 times heavier than electrons.
- The total number of protons in the atoms of an element is always equal to the atomic number of the element.

Neutrons

- The mass of a neutron is almost the same as that of a proton i.e. 1.674×10^{-24}.
- Neutrons are electrically neutral particles and carry no charge.
- Different isotopes of an element have the same number of protons but vary in the number of neutrons present in their respective nuclei.

Electrons

- The charge of an electron is -1e, which approximates to -1.602×10^{-19}.
- The mass of an electron is approximately 9.1×10^{-31}.
- Due to the relatively negligible mass of electrons, they are ignored when calculating the mass of an atom.

Atomic Structure of Isotopes

Nucleons are the components of the nucleus of an atom. A nucleon can either be a proton or a neutron. Each element has a unique number of protons in it, which is described by its unique atomic number. However, several atomic structures of an element can exist, which differ in the total number of nucleons.

These variants of elements having a different nucleon number (also known as the mass number) are called isotopes of the element. Therefore, the isotopes of an element have the same number of protons but differ in the number of neutrons.

The atomic structure of an isotope is described with the help of the chemical symbol of the element, the atomic number of the element, and the mass number of the isotope. For example, there exist three known naturally occurring isotopes of hydrogen, namely, protium, deuterium, and tritium.

The atomic structures of these hydrogen isotopes are illustrated below.

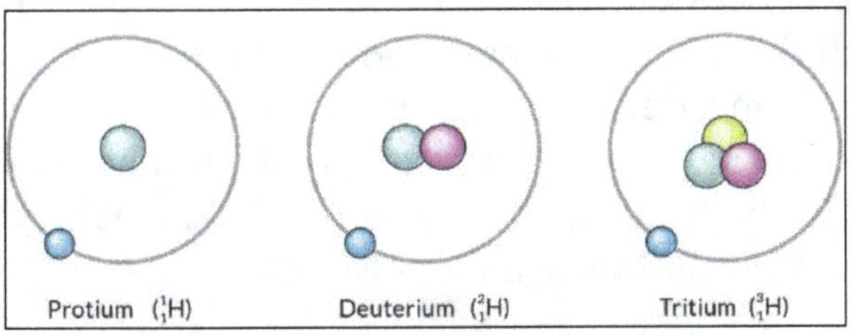

Atomic structures of hydrogen isotopes.

The isotopes of an element vary in stability. The half-lives of isotopes also differ. However, they generally have similar chemical behavior owing to the fact that they hold the same electronic structures.

Crystal Symmetry

crystal symmetry In well-formed crystals, the symmetrically arranged faces reflect the internal arrangement of atoms. The symmetry of individual crystals is determined by reference to three elements. The plane of symmetry (also called the 'mirror plane' or 'symmetry plane') is a plane by which the crystal may be divided into two halves which are mirror images of each other. The axis of symmetry is a line about which a crystal may be rotated through 360°/n until it assumes a congruent position; n may equal 2, 3, 4, or 6 (but not 1), depending on the number of times the congruent position is repeated. These correspond respectively to 2-fold (diad), 3-fold (triad), 4-fold (tetrad), and 6-fold (hexad) axes. The centre of symmetry is a central point which is present when all faces or edges occur in parallel pairs on opposite sides of the crystal. Using these elements of symmetry, crystallographers have recognized 32 crystal classes and seven crystal systems. Symmetry is highest (high symmetry) in the cubic system, where many elements are repeated, and lowest (low symmetry) in the triclinic system, where only a centre of symmetry may be present (i.e. there may be no plane or axis of symmetry).

Miller Indices

Miller indices are group of three numbers that indicates the orientation of a plane or set of parallel planes of atoms in a crystal. If each atom in the crystal is represented by a point and these points are connected by lines, the resulting lattice may be divided into a number of identical blocks, or unit cells; the intersecting edges of one of the unit cells defines a set of crystallographic axes, and the Miller indices are determined by the intersection of the plane with these axes. The reciprocals of these intercepts are computed, and fractions are cleared to give the three Miller indices (hkl). For example, a plane parallel to two axes but cutting the third axis at a length equal to one edge of a unit cell has Miller indices of (100), (010), or (001), depending upon the axis cut; and a plane cutting all three axes at lengths equal to the edges of a unit cell has Miller indices of (111). This scheme, devised by British mineralogist and crystallographer William Hallowes Miller, in 1839, has the advantage of eliminating all fractions from the notation for a plane. In the hexagonal system, which has four crystallographic axes, a similar scheme of four Bravais-Miller indices is used.

Planes in crystals are described by a set of integers (h, k and l) known as Miller indices. Miller indices of a plane are the reciprocals of the fractional intercepts of that plane on the various crystallographic axes. For calculating Miller indices, a reference plane, known as parametral plane, is selected having intercepts a, b and c along x, y and z-axes, respectively. Then, the intercepts of the unknown plane are given with respect to a, b and c of the parametral plane.

Thus, the Miller indices are:

$$h = \frac{a}{intercept\ of\ the\ plane\ along\ x-axis}$$

$$k = \frac{b}{intercept\ of\ the\ plane\ along\ y-axis}$$

$$l = \frac{c}{intercept\ of\ the\ plane\ along\ z-axis}$$

The distance between the parallel planes in crystals are designated as d_{hkl}. For different cubic lattices these interplanar spacing are given by the general formula,

$$d_{(hkl)} = \frac{a}{\sqrt{h^2 + k^2 + l^2}}$$

where a is the length of cube side while h, k and l are the Miller indices of the plane.

When a plane is parallel to an axis, its intercept with that axis is taken as infinite and the Miller will be zero.

Negative signs in the Miller indices is indicated by placing a bar on the intercept. All parallel planes have same Miller indices.

Crystal Defects

Crystal defect is imperfection in the regular geometrical arrangement of the atoms in a crystalline solid. These imperfections result from deformation of the solid, rapid cooling from high temperature, or high-energy radiation (X-rays or neutrons) striking the solid. Located at single points, along lines, or on whole surfaces in the solid, these defects influence its mechanical, electrical, and optical behavior.

Point Defects

Departures of a crystalline solid from a regular array of atoms or ions. A "perfect" crystal of NaCl, for example, would consist of alternating Na^+ and Cl^- ions on an infinite three-dimensional simple cubic lattice, and a simple defect (a vacancy) would be a missing Na^+ or Cl^- ion. There are many other kinds of possible defects, ranging from simple and microscopic, such as the vacancy and other structures shown in the illustration, to complex and macroscopic, such as the inclusion of another material, or a surface.

Natural crystals always contain defects, due to the uncontrolled conditions under which they were formed. The presence of defects which affect the color can make these crystals valuable as gems, as in ruby (Cr replacing a small fraction of the Al in Al_2O_3). Crystals prepared in the laboratory will also always contain defects, although considerable control may be exercised over their type, concentration, and distribution.

The importance of defects depends upon the material, type of defect, and properties which are being considered. Some properties, such as density and elastic constants, are proportional to the concentration of defects, and so a small defect concentration will have a very small effect on these. Other properties, such as the conductivity of a semiconductor crystal, may be much more sensitive to the presence of small numbers of defects. Indeed, while the term defect carries with it the connotation of undesirable qualities, defects are responsible for many of the important properties of materials, and much of solid-state physics and materials science involves the study and engineering of defects so that solids will have desired properties. A defect-free silicon crystal would be of little use in modern electronics; the use of silicon in devices is dependent upon small concentrations of chemical impurities such as phosphorus and arsenic which give it desired electronic properties.

An important class of crystal defect is the chemical impurity. The simplest case is the substitutional impurity, for example, a zinc atom in place of a copper atom in metallic copper. Impurities may also be interstitial; that is, they may be located where atoms or ions normally do not exist. In metals, impurities usually lead to an increase in the electrical resistivity. Impurities in semiconductors are responsible for the important electrical properties which lead to their widespread use. The energy levels associated with impurities and other defects in nonmetals may also lead to optical absorption in interesting regions of the spectrum.

Even in a chemically pure crystal, structural defects will occur. These may be simple or extended. One type of simple defect is the vacancy, but other types exist. The atom which left a normal site to create a vacancy may end up in an interstitial position, a location not normally occupied. Or it may form a bond with a normal atom in such a way that neither atom is on the normal site, but the two are symmetrically displaced from it. This is called a split interstitial. The name Frenkel defect is given to a vacancy-interstitial pair, whereas an isolated vacancy is a Schottky defect.

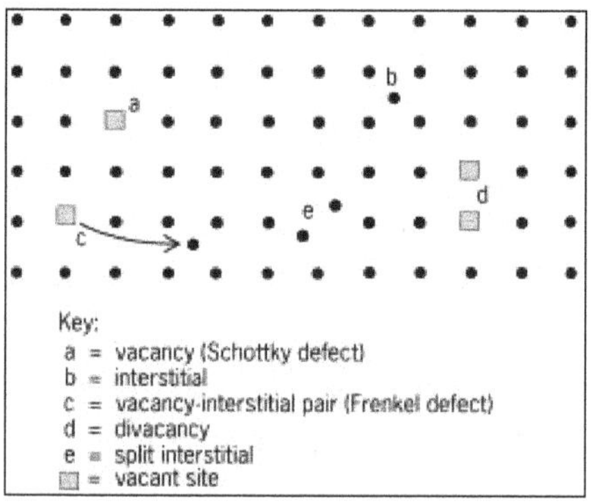

Some simple defects in a lattice.

The simplest extended structural defect is the dislocation. An edge dislocation is a line defect which may be thought of as the result of adding or subtracting a half-plane of atoms. A screw dislocation is a line defect which can be thought of as the result of cutting partway through the crystal and displacing it parallel to the edge of the cut. Dislocations are of great importance in determining the mechanical properties of crystals. A dislocation-free crystal is resistant to shear, because atoms must be displaced over high-potential-energy barriers from one equilibrium position to another. It takes relatively little energy to move a dislocation (and thereby shear the crystal), because the atoms at the dislocation are barely in stable equilibrium. Such plastic deformation is known as slip.

For both scientific and practical reasons, much of the research on crystal defects is directed toward the dynamic properties of defects under particular conditions, or defect chemistry. Much of the motivation for this arises from the often undesirable effects of external influences on material properties, and a desire to minimize these effects. Examples of defect chemistry abound, including one as familiar as the photographic process, in which incident photons cause defect modifications in silver halides or other materials. Properties of materials in nuclear reactors is another important case.

Disruptions in the regularity of the crystal structure in real single crystals. In idealized crystal structures, atoms occupy strictly defined positions, forming regular three-dimensional lattices (crystal lattices). In natural and artificially grown real crystals, various deviations from the regular location of atoms or ions or of groups of atoms or ions are usually observed. Such disturbances may be either on an atomic scale or of macroscopic dimensions, noticeable even to the naked eye. In addition to static defects, there exists another type of deviations from the ideal lattice that are related to thermal oscillations of the particles that make up the lattice (dynamic defects).

Crystal defects are formed during growth of crystals; under the influence of thermal, mechanical, and electrical factors; and upon irradiation by neutrons, electrons, X rays, and ultraviolet radiation (radiation defects).

A distinction is made among point (zero-dimensional) defects, line (one-dimensional) defects, defects that form surfaces in the crystal (two-dimensional), and volumetric (three-dimensional) defects. In a one-dimensional defect the size is much greater in one direction than the distance between neighboring like atoms (the lattice parameter), but in the other two directions it is of the same order. In a two-dimensional defect the dimensions are greater in two directions than is the distance between the closest atoms, and so on.

Stoichiometric Defects

The compounds in which the number of positive and negative ions are exactly in the ratios indicated by their chemical formulae are called stoichiometric compounds. The defects do not disturb the stoichiometry (the ratio of numbers of positive and negative ions) are called stoichiometric defects. These are of following types:

Vacancy Defect

In crystallography, a vacancy is a type of point defect in a crystal. Crystals inherently possess imperfections, sometimes referred to as crystalline defects. A defect in which an atom is missing from one of the lattice sites is known as a “vacancy” defect.

It is also known as a Schottky defect, although in ionic crystals the concepts are not identical.

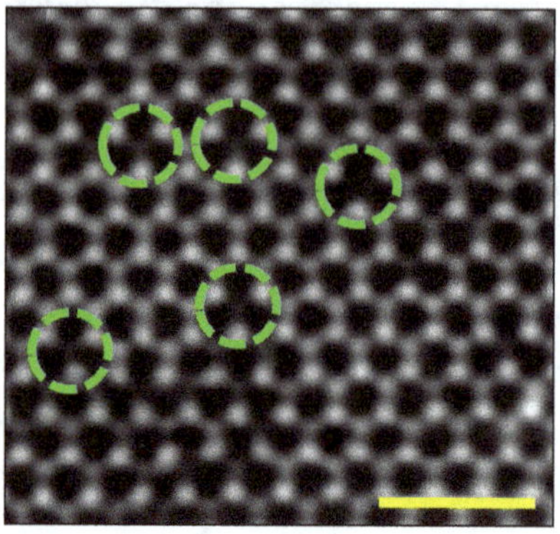

Electron microscopy of sulfur vacancies in a monolayer of molybdenum disulfide. Right circle points to a divacancy, i.e., sulfur atoms are missing both above and below the Mo layer. Other circles are single vacancies, i.e., sulfur atoms are missing only above or below the Mo layer. Scale bar: 1 nm.

Vacancies occur naturally in all crystalline materials. At any given temperature, up to the melting point of the material, there is an equilibrium concentration (ratio of vacant lattice sites to those containing atoms). At the melting point of some metals the ratio can be approximately 1:1000. This temperature dependence can be modelled by:

$$N_V = N \exp(-Q_v/k_B T)$$

where N_v is the vacancy concentration, Q_v is the energy required for vacancy formation, k_B is the Boltzmann constant, T is the absolute temperature, and N is the concentration of atomic sites i.e.

$$N = pN_A/\mathrm{A}$$

where ρ is density, N_A Avogadro constant, and A the atomic mass.

It is the simplest point defect. In this system, an atom is missing from its regular atomic site. Vacancies are formed during solidification due to vibration of atoms, local rearrangement of atoms, plastic deformation and ionic bombardments.

The creation of a vacancy can be simply modeled by considering the energy required to break the bonds between an atom inside the crystal and its nearest neighbor atoms. Once that atom is removed from the lattice site, it is put back on the surface of the crystal and some energy is retrieved because new bonds are established with other atoms on the surface. However, there is a net input of energy because there are fewer bonds between surface atoms than between atoms in the interior of the crystal.

Schottky Defect

A Schottky defect is a type of point defect in a crystal lattice named after Walter H. Schottky. In non-ionic crystals it refers to lattice valency defects.

In ionic crystals, this type of point defect forms when oppositely charged ions leave their lattice sites, creating vacancies. These vacancies are formed in stoichiometric units, to maintain an overall neutral charge in the ionic solid. The surrounding atoms then move to fill these vacancies, causing new vacancies to form. Normally these defects will lead to a decrease in the density of the crystal or metal. Chemical equations in Kröger–Vink notation for the formation of Schottky defects in TiO_2 and $BaTiO_3$.

$$\theta \rightleftharpoons v_{Ti}^{''''} + 2v_{O}^{\bullet\bullet}$$

$$\theta \rightleftharpoons v_{Ba}^{''} + v_{Ti}^{''''} + 3v_{O}^{\bullet\bullet}$$

This can be illustrated schematically with a two-dimensional diagram of a sodium chloride crystal lattice:

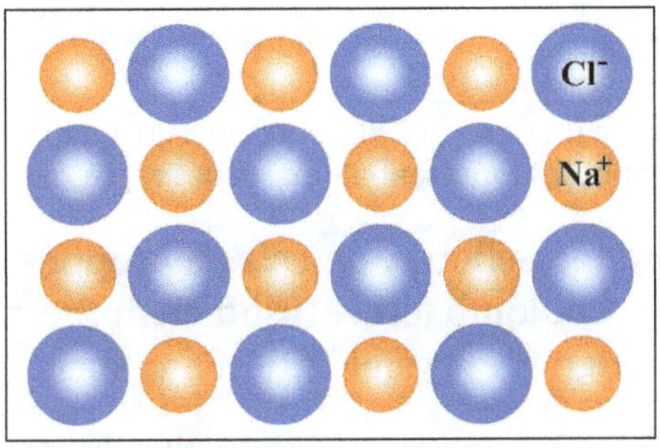

The defect-free NaCl structure.

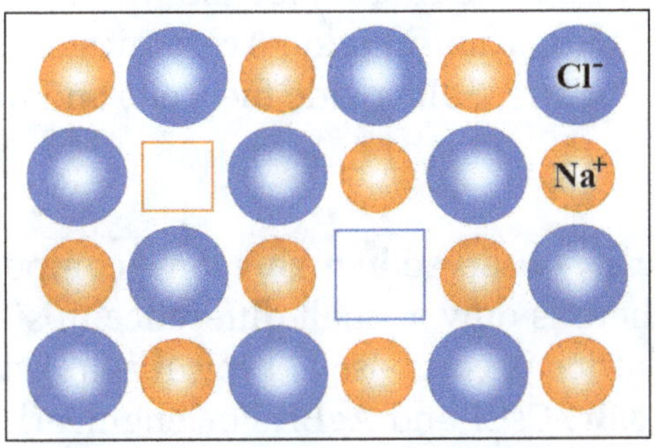

Schottky defects within the NaCl structure.

Schottky defects consist of unoccupied anion and cation sites in a stoichiometric ratio. For a simple ionic crystal of type A^-B^+, a Schottky defect consists of a single anion vacancy (A) and a single cation vacancy (B), or B), or $v_A^{\bullet} + v_B^{'}$ following Kröger–Vink notation. For a more general crystal with formula A_xB_y, a Schottky cluster is formed of x vacancies of A and y vacancies of B, thus the overall stoichiometry and charge neutrality are conserved.

Schottky defects are observed most frequently when there is a small difference in size between the cations and anions that make up a material. They can be produced as the result of the thermal incorporation of unoccupied lattice sites from the exterior of the crystal. The lattice undergoes thermal vibration and thermal expansion when the temperature is raised. When this happens a pair of vacancies is incorporated in the crystal and charge neutrality is maintained. The ion which is missing from the lattice point occupies an interstitial position.

Bound and Dilute Defects

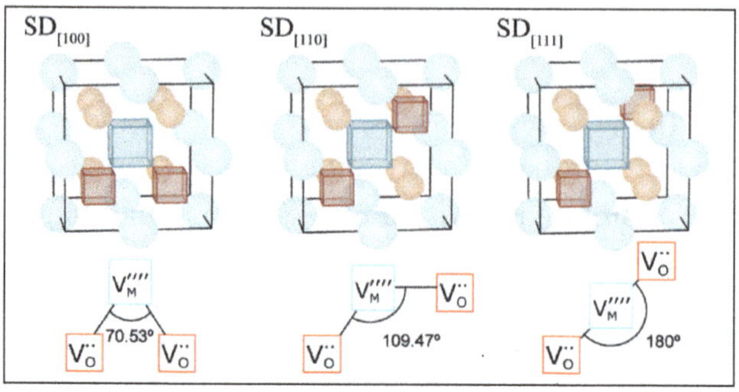

Three bound configurations of Schottky defects in an oxide with Fluorite structure. Spheres represent atoms, cubes represent vacancies.

The vacancies that make up the Schottky defects have opposite charge, thus they experience a mutually attractive Coulomb force. If sufficient thermal energy is available the vacancies may migrate through the crystal lattice, and form bound clusters.

The bound clusters are typically less mobile than the dilute counterparts, as multiple species need to move in a concerted motion for the whole cluster to migrate. This has important implications for numerous functional ceramics used in a wide range of applications, including ion conductors, Solid oxide fuel cells and nuclear fuel.

Examples:

This type of defect is typically observed in highly ionic compounds, highly coordinated compounds, and where there is only a small difference in sizes of cations and anions of which the compound lattice is composed. Typical salts where Schottky disorder is observed are NaCl, KCl, KBr, CsCl and AgBr. For engineering applications, Schottky defects are important in oxides with Fluorite structure, such as CeO_2, cubic ZrO_2, UO_2, ThO_2 and PuO_2.

Effect on Density

Since the total number of ions present in the crystal with this defect is less than the theoretical number of ions for a crystal of its volume, the density of the solid crystal is less than the theoretical density of the material.

Frenkel Defect

A Frenkel defect or dislocation defect is a type of point defect in crystalline solids named after its discoverer Yakov Frenkel. The defect forms when an atom or smaller ion (usually cation) leaves its place in the lattice, creating a vacancy, and becomes an interstitial by lodging in a nearby location. Their primary mechanism of generation is by particle irradiation, as their equilibrium concentration according to the Boltzmann distribution is much smaller than the pure interstitial atoms.

In an ionic crystal, this point defect forms when an ion is displaced from its lattice position to an interstitial site, creating a vacancy at the original site and an interstitial defect at the new location without any changes in chemical properties. This defect appears in an ionic solid which usually possess low coordination number or a considerable disparity in the sizes of the ions. The smaller ion (usually the cation) is dislocated; e.g. in ZnS, AgCl, AgBr, and AgI due to the small size of the zinc ion with charge +2 and the silver ion with charge +1, the cations in the respective cases get dislocated.

AgBr exhibits both Frenkel defect and Schottky defect despite the disparity in the sizes of the constituent ions.

Effect on Density

Frenkel defect does not have any impact on the density of the solid as it involves only the migration of the ions within the crystal, thus preserving both the volume as well as mass.

Examples:

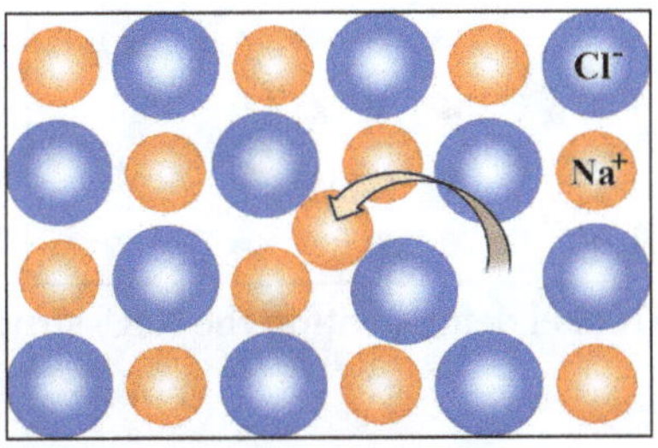

The Frenkel defect within the NaCl structure.

Frenkel defects are exhibited in ionic solids with a large size difference between the anion and cation (with the cation usually smaller due to an increased effective nuclear charge).

Some examples of solids which exhibit Frenkel defects:

- Zinc sulfide.
- Silver(I) chloride.
- Silver(I) bromide (also shows Schottky defects).
- Silver(I) iodide.

These are due to the comparatively smaller size of Zn^{2+} and Ag^{+} ions:

For example, consider a lattice formed by X^{n-} and M^{n+} ions. Suppose an M ion leaves the M sublattice, leaving the X sublattice unchanged. The number of interstitials formed will equal the number of vacancies formed.

One form of a Frenkel defect reaction in MgO with the oxide anion leaving the lattice and going into the interstitial site written in Kröger–Vink notation:

$$Mg_{Mg}^{\times} + O_{o}^{\times} \rightarrow O_{i}^{\prime\prime} + v_{o}^{\bullet\bullet} + Mg_{Mg}^{\times}$$

This can be illustrated with the example of the sodium chloride crystal structure. The diagrams below are schematic two-dimensional representations.

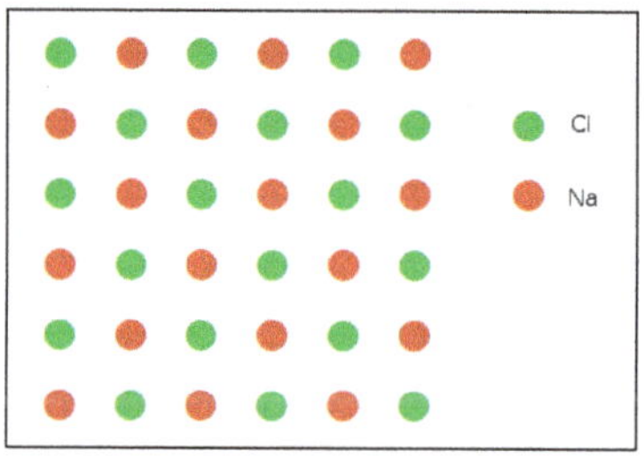

The defect-free NaCl structure.

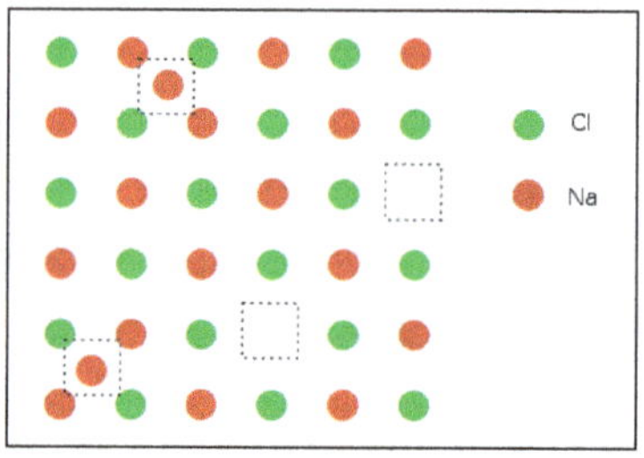

Two Frenkel defects within the NaCl structure.

Interstitial Defect

Interstitials defects are a variety of crystallographic defects where atoms assume a normally unoccupied site in the crystal structure. In interstitial defects three or more atoms may share one lattice site, thereby increasing its total energy. Alternatively small atoms in some crystals may occupy interstitial sites in energetically favorable configurations, such as hydrogen in palladium. Interstitials can be produced by bombarding a crystal with elementary particles having energy above the displacement threshold for that crystal, but they may also exist in small concentrations in thermodynamic equilibrium.

Self-interstitials

Self-interstitial defects are interstitial defects which contain only atoms which are the same as those already present in the lattice.

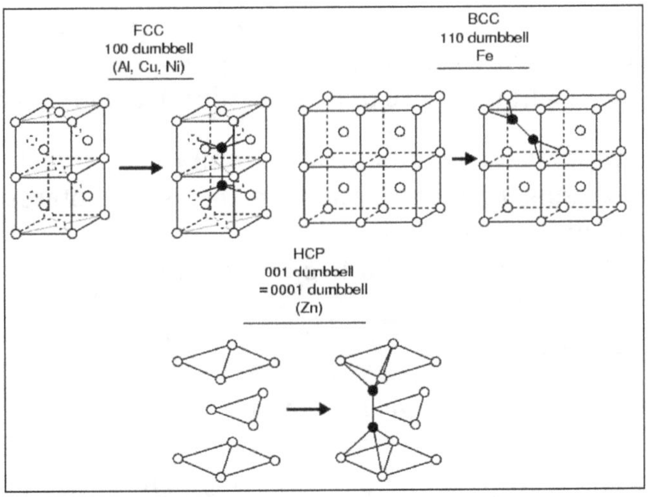

Structure of self-interstitial in some common metals. The left-hand side of each crystal type shows the perfect crystal and the right-hand side the one with a defect.

The structure of interstitial defects has been experimentally determined in some metals and semiconductors.

Contrary to what one might intuitively expect, most self-interstitials in metals with a known structure have a 'split' structure, in which two atoms share the same lattice site. Typically the center of mass of the two atoms is at the lattice site, and they are displaced symmetrically from it along one of the principal lattice directions. For instance, in several common face-centered cubic (fcc) metals such as copper, nickel and platinum, the ground state structure of the self-interstitial is the split interstitial structure, where two atoms are displaced in a positive and negative direction from the lattice site. In body-centered cubic (bcc) iron the ground state interstitial structure is similarly a split interstitial.

These split interstitials are often called dumbbell interstitials, because plotting the two atoms forming the interstitial with two large spheres and a thick line joining them makes the structure resemble a dumbbell weight-lifting device.

In other bcc metals than iron, the ground state structure is believed based on recent density-functional theory calculations to be the crowdion interstitial, which can be understood as a long chain (typically some 10–20) of atoms along the lattice direction, compressed compared to the perfect lattice such that the chain contains one extra atom.

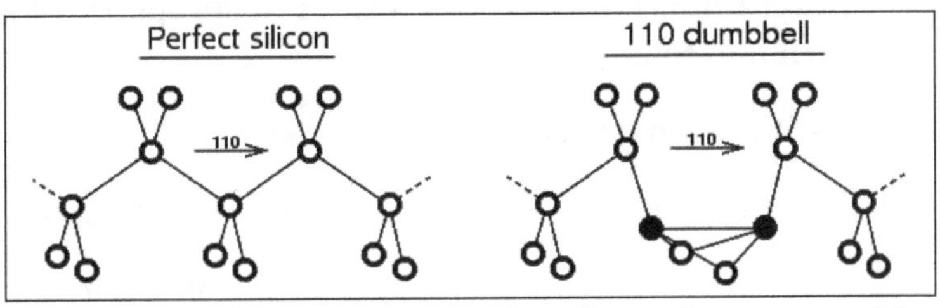

Structure of dumbbell self-interstitial in silicon. The structure of the interstitial in silicon may depend on charge state and doping level of the material.

In semiconductors the situation is more complex, since defects may be charged and different charge states may have different structures. For instance, in silicon, the interstitial may either have a split structure or a tetrahedral truly interstitial one.

Carbon, notably in graphite and diamond, has a number of interesting self-interstitials - recently discovered using LDF-calculations is the "spiro-interestitial" in graphite, named after spiropentane, as the interstitial carbon atom is situated between two basal planes and bonded in a geometry similar to spiropentane.

Impurity Interstitials

Small impurity interstitial atoms are usually on true off-lattice sites between the lattice atoms. Such sites can be characterized by the symmetry of the interstitial atom position with respect to its nearest lattice atoms. For instance, an impurity atom I with 4 nearest lattice atom A neighbours (at equal distances) in an fcc lattice is in a tetrahedral symmetry position, and thus can be called a tetrahedral interstitial.

Large impurity interstitials can also be in split interstitial configurations together with a lattice atom, similar to those of the self-interstitial atom.

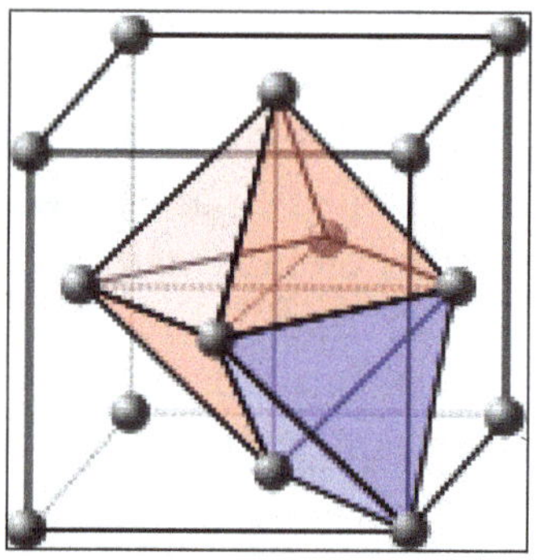

Octahedral (red) and tetrahedral (blue) interstitial symmetry polyhedra in a face-centered cubic lattice. The actual interstitial atom would ideally be in the middle of one of the polyhedra.

Effects of Interstitials

Interstitials modify the physical and chemical properties of materials:

- Interstitial carbon atoms have a crucial role for the properties and processing of steels, in particular carbon steels.
- Impurity interstitials can be used e.g. for storage of hydrogen in metals.
- The crystall lattice can expand with the concentration of impurity interstitials.
- The amorphization of semiconductors such as silicon during ion irradiation is often explained by the buildup of a high concentration of interstitials leading eventually to the collapse of the lattice as it becomes unstable.

- Creation of large amounts of interstitials in a solid can lead to a significant energy buildup, which on release can even lead to severe accidents in certain old types of nuclear reactors (Wigner effect). The high-energy states can be released by annealing.
- At least in fcc lattice, interstitials have a large diaelastic softening effect on the material.
- It has been proposed that interstitials are related to the onset of melting and the glass transition.

Stone-Wales Defect

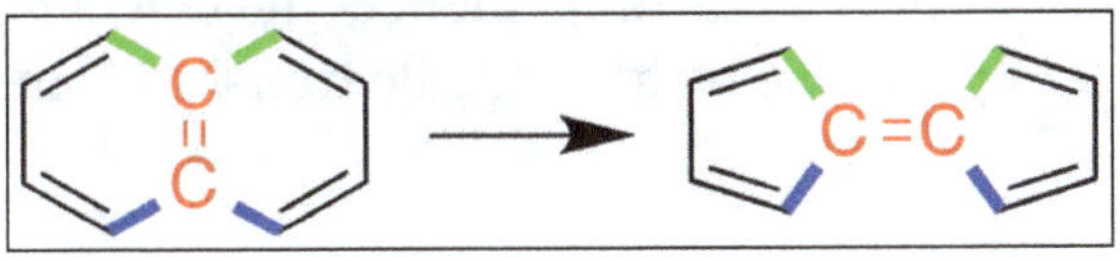

A Stone–Wales defect is a crystallographic defect that involves the change of connectivity of two π-bonded carbon atoms, leading to their rotation by 90° with respect to the midpoint of their bond. The reaction commonly involves conversion between a naphthalene-like structure into a fulvalene-like structure, that is, two rings that share an edge vs two separate rings that have vertices bonded to each other.

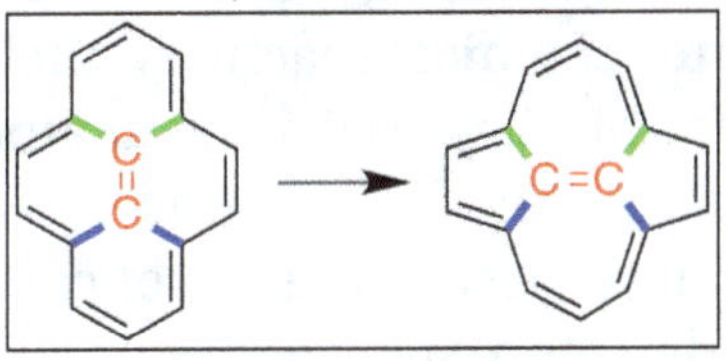

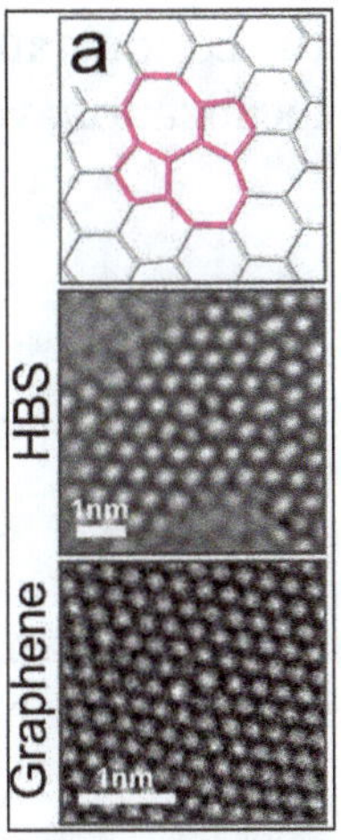

Stone–Wales defect in 2D silica(HBS, middle) and graphene (bottom): model and TEM images.

The reaction occurs on carbon nanotubes, graphene, and similar carbon frameworks, where the four adjacent six-membered rings of a pyrene-like region are changed into two five-membered rings and two seven-membered rings when the bond uniting two of the adjacent rings rotates. In these materials, the rearrangement is thought to have important implications for the chemical, electrical, and mechanical properties. The rearrangement is an example of a pyracyclene rearrangement.

Structural Effects

The defects have been imaged using scanning tunneling microscopy and transmission electron microscopy and can be determined using various vibrational spectroscopy-techniques.

It has been proposed that the coalescence process of fullerenes or carbon nanotubes may occur through a sequence of such a rearrangements. The defect is thought to be responsible for nanoscale plasticity and the brittle–ductile transitions in carbon nanotubes.

Chemical Details

The activation energy for the simple atomic motion that gives the bond-rotation apparent in a Stone–Wales defects is fairly high—a barrier of several electronvolts. But various processes can create the defects at substantially lower energies than might be expected.

The rearrangement creates a structure with less resonance stabilization among the sp^2 atoms involved and higher strain energy in the local structure. As a result, the defect creates a region with greater chemical reactivity, including acting as a nucleophile and creating a preferred site for binding to hydrogen atoms. The high affinity of these defects for hydrogen, coupled with the large surface area of the bulk material, might make these defects an important aspect in the use of carbon nanomaterials for hydrogen storage. Incorporation of defects along a carbon-nanotube network can program a carbon-nanotube circuit to enhance the conductance along a specific path. In this scenario, the defects lead to a charge delocalization, which redirects an incoming electron down a given trajectory.

Non-Stoichiometric Defects

These defects are of two types:

Metal Excess Defect

Metal excess defect due to anionic vacancies: In this case, negative ions may be missing from their lattice sites leaving holes in which the electrons remain entrapped to maintain the electrical neutrality.

This type of defect is observed in those crystals which are likely to form Schottky Defects. For example, Alkali metal halides like NaCl and KCl show this type of defect. When crystals of NaCl are heated in an atmosphere of sodium vapour, the sodium atoms are deposited on the surface of the crystal. The Cl^- ions diffuse to the surface of the crystal and combine with Na atoms to give NaCl. This happens by loss of electron by sodium atoms to form Na^+ ions. The released electrons diffuse into the crystal and occupy anionic sites. As a result the crystal now has an excess of sodium. The anionic sites occupied by unpaired electrons are called *F-centres* (from the German word *Farbenzenter* for colour centre). They impart yellow colour to the crystals of NaCl. The colour results by excitation of these electrons when they absorb energy from the visible light falling on the crystals. Similarly, excess of lithium makes LiCl crystals pink and excess of potassium makes KCl crystals violet (or lilac).

Metal excess defect due to the presence of extra cations at interstitial sites: In this case, there are extra positive ions occupying interstitial sites and the electrons in another interstitial sites to maintain electrical neutrality. The defect may be visualised as the loss of non-metal atoms which leave their electrons behind. The excess metal ions occupy interstitial positions.

The common example is Zinc Oxide. Zinc oxide is white in colour at room temperature. On heating it loses oxygen and turns yellow:

$$ZnO \rightarrow Zn^{2+} + \tfrac{1}{2}O_2 + 2e^-$$

Now there is excess of zinc in the crystal and its formula becomes $Zn_{1+x}O$. The excess Zn^{2+} ions move to interstitial sites and the electrons to neighbouring interstitial sites.

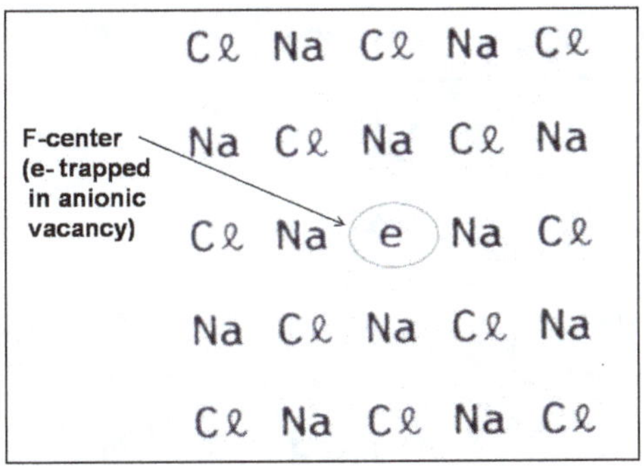

- ZnO heated in Zn vapour $\rightarrow Zn_yO$ $(y > 1)$.
- The excess cations occupy interestial voids.
- The electrons ($2e^-$) released stay associated to the interestial cations.

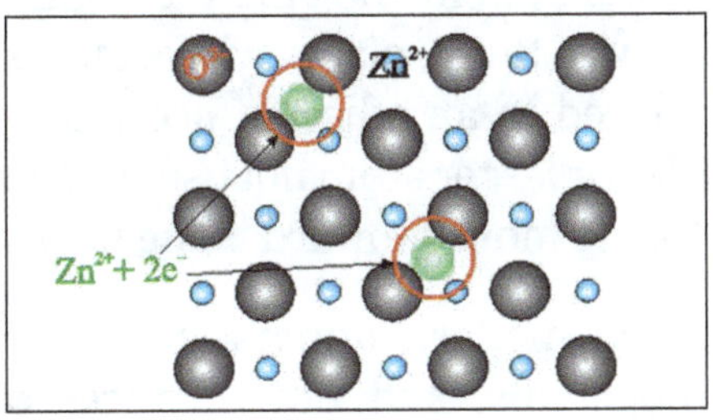

Metal Deficiency Defects

These contain less number of positive ions than negative ions. These arise due to two ways:

- Cation Vacancies: In some cases, the positive ions may be missing from their lattice sites. The extra negative charge may be balanced by some nearby metal ion acquiring two positive charges instead of one. This type of defect is possible in metals which show variable oxidation states.

The common examples of compounds having this defect are ferrous oxide, ferrous sulphide, nickel oxide etc.

- Extra anions occupying interstitial sites: In the case, the extra anions may be occupying interstitial positions. The extra negative charge is balanced by the extra charges on the adjacent metal ions. Such type of defect is not common because the negative ions usually very large and they cannot easily fit into the interstitial sites.

Non-stoichiometric Compound

Non-stoichiometric compounds are chemical compounds, almost always solid inorganic compounds, having elemental composition whose proportions cannot be represented by a ratio of small natural numbers; most often, in such materials, some small percentage of atoms are missing or too many atoms are packed into an otherwise perfect lattice work.

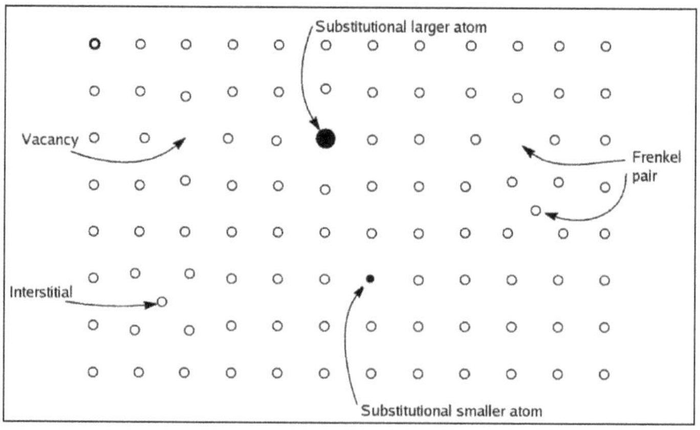

Origin of title phenomena in crystallographic defects. Shown is a two-dimensional slice through a primitive cubic crystal system showing the regular square array of atoms on one face (open circles, o), and with these, places where atoms are missing from a regular site to create vacancies, displaced to an adjacent acceptable space to create a Frenkel pair, or substituted by a smaller or larger atom not usually seen (closed circles, •), in each case resulting in a material that is moved toward being measurably non-stoichiometric.

Contrary to earlier definitions, modern understanding of non-stoichiometric compounds view them as homogenous, and not mixtures of stoichiometric chemical compounds. Since the solids are overall electrically neutral, the defect is compensated by a change in the charge of other atoms in the solid, either by changing their oxidation state, or by replacing them with atoms of different elements with a different charge. Many metal oxides and sulfides have non-stoichiometric examples; for example, stoichiometric iron(II) oxide, which is rare, has the formula FeO, whereas the more common material is nonstoichiometric, with the formula $Fe_{0.95}O$. The type of equilibrium defects in non-stoichiometric compounds can vary with attendant variation in bulk properties of the material. Non-stoichiometric compounds also exhibit special electrical or chemical properties because of the defects; for example, when atoms are missing, electrons can move through the solid more rapidly. Non-stoichiometric compounds have applications in ceramic and superconductive material and in electrochemical (i.e., battery) system designs.

Occurrence

Iron Oxides

Nonstoichiometry is pervasive for metal oxides, especially when the metal is not in its highest oxidation state. For example, although wüstite (ferrous oxide) has an ideal (stoichiometric) formula FeO, the actual stoichiometry is closer to $Fe_{0.95}O$. The non-stoichiometry reflect the ease of oxidation of Fe^{2+} to Fe^{3+} effectively replacing a small portion of Fe^{2+} with two thirds their number of Fe^{3+}. Thus for every three "missing" Fe^{2+} ions, the crystal contains two Fe^{3+} ions to balance the charge. The composition of a non-stoichiometric compound usually varies in a continuous manner over a narrow range. Thus, the formula for wüstite is written as $Fe_{1-x}O$, where x is a small number (0.05 in the previous example) representing the deviation from the "ideal" formula. Nonstoichiometry is especially important in solid, three-dimensional polymers that can tolerate mistakes. To some extent, entropy drives all solids to be non-stoichiometric. But for practical purposes, the term describes materials where the non-stoichiometry is measurable, usually at least 1% of the ideal composition.

Iron Sulfides

The monosulfides of the transition metals are often nonstoichiometric. Best known perhaps is nominally iron(II) sulfide (the mineral pyrrhotite) with a composition $Fe_{1-x}S$ (x = 0 to 0.2). The rare stoichiometric FeS endmember is known as the mineral *troilite*. Pyrrhotite is remarkable in that it has numerous polytypes, i.e. crystalline forms differing in symmetry (monoclinic or hexagonal) and composition (Fe_7S_8, Fe_9S_{10}, $Fe_{11}S_{12}$ and others). These materials are always iron-deficient owing to the presence of lattice defects, namely iron vacancies. Despite those defects, the composition is usually expressed as a ratio of large numbers and the crystals symmetry is relatively high. This means the iron vacancies are not randomly scattered over the crystal, but form certain regular configurations. Those vacancies strongly affect the magnetic properties of pyrrhotite: the magnetism increases with the concentration of vacancies and is absent for the stoichiometric FeS.

Pyrrhotite, an example of a non-stoichiometric inorganic compound, with formula $Fe_{1-x}S$ (x = 0 to 0.2).

Palladium Hydrides

Palladium hydride is a nonstoichiometric material of the approximate composition PdH_x ($0.02 < x < 0.58$). This solid conducts hydrogen by virtue of the mobility of the hydrogen atoms within the solid.

Tungsten Oxides

It is sometimes difficult to determine if a material is non-stoichiometric or if the formula is best represented by large numbers. The oxides of tungsten illustrate this situation. Starting from the idealized material tungsten trioxide, one can generate a series of related materials that are slightly deficient in oxygen. These oxygen-deficient species can be described as WO_{3-x}, but in fact they are stoichiometric species with large unit cells with the formulas W_nO_{3n-2}, where n = 20, 24, 25, 40. Thus, the last species can be described with the stoichiometric formula $W_{40}O_{118}$, whereas the non-stoichiometric description $WO_{2.95}$ implies a more random distribution of oxide vacancies.

Other Cases

At high temperatures (1000 °C), titanium sulfides present a series of non-stoichiometric compounds.

The coordination polymer Prussian blue, nominally $Fe_7(CN)_{18}$ and their analogs are well known to form in non-stoichiometric proportions. The non-stoichiometric phases exhibit useful properties vis-à-vis their ability to bind caesium and thallium ions.

Applications

Oxidation catalysis

Many useful compounds are produced by the reactions of hydrocarbons with oxygen, a conversion that is catalyzed by metal oxides. The process operates via the transfer of "lattice" oxygen to the hydrocarbon substrate, a step that temporarily generates a vacancy (or defect). In a subsequent step, the missing oxygen is replenished by O_2. Such catalysts rely on the ability of the metal oxide to form phases that are not stoichiometric. An analogous sequence of events describes other kinds of atom-transfer reactions including hydrogenation and hydrodesulfurization catalysed by solid catalysts. These considerations also highlight the fact that stoichiometry is determined by the interior of crystals: the surfaces of crystals often do not follow the stoichiometry of the bulk. The complex structures on surfaces are described by the term "surface reconstruction".

Ion Conduction

The migration of atoms within a solid is strongly influenced by the defects associated with non-stoichiometry. These defect sites provide pathways for atoms and ions to migrate through the otherwise dense ensemble of atoms that form the crystals. Oxygen sensors and solid state batteries are two applications that rely on oxide vacancies. One example is the CeO_2-based sensor in automotive exhaust systems. At low partial pressures of O_2, the sensor allows the introduction of increased air to effect more thorough combustion.

Superconductivity

Many superconductors are non-stoichiometric. For example, yttrium barium copper oxide, arguably the most notable high-temperature superconductor, is a non-stoichiometric solid with the formula $Y_xBa_2Cu_3O_{7-x}$. The critical temperature of the superconductor depends on the exact value of x. The stoichiometric species has $x = 0$, but this value can be as great as 1.

Impurity Defects

These defects arise when foreign atoms are present at the lattice site (in place of host atoms) or at the vacant interstitial sites. In the former case, we get substitutional solid solutions while in the latter case, we get interstitial solid solution. The formation of the former depends upon the electronic structure of the impurity while that of the later on the size of the impurity.

Line Defects

Line defect is an imperfection in crystal structure characterized by an atomic layer extending part way into a crystal structure.

Dislocation

In materials science, a dislocation or Taylor's dislocation is a linear crystallographic defect or irregularity within a crystal structure which contains an abrupt change in the arrangement of atoms. The movement of dislocations allow atoms to slide over each other at low stress levels and is known as *glide* or slip. The crystalline order is restored on either side of a *glide dislocation* but the atoms on one side have moved by one position. The crystalline order is not fully restored with a *partial dislocation*. Dislocations define the boundary between *slipped* and *unslipped* regions of material. A dislocation can be characterised by the distance and direction of movement it causes to atoms which is defined by the Burgers vector. Plastic deformation of a material occurs by the creation and movement of many dislocations. The number and arrangement of dislocations influences many of the properties of materials.

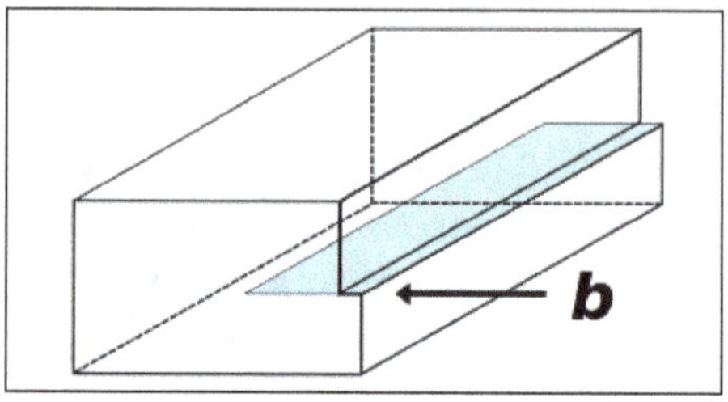

An edge-dislocation (b = Burgers vector).

The two primary types of dislocations are *sessile* dislocations which are immobile and *glissile* dislocations which are mobile. Examples of sessile dislocations are the *stair-rod* dislocation and the Lomer–Cottrell junction. The two main types of mobile dislocations are *edge* and *screw* dislocations.

Edge dislocations can be visualized as being caused by the termination of a plane of atoms in the middle of a crystal. In such a case, the surrounding planes are not straight, but instead bend around the edge of the terminating plane so that the crystal structure is perfectly ordered on either side. This phenomenon is analogous to half of a piece of paper inserted into a stack of paper, where the defect in the stack is noticeable only at the edge of the half sheet.

The theory describing the elastic fields of the defects was originally developed by Vito Volterra in 1907. In 1934, Egon Orowan, Michael Polanyi and G. I. Taylor, proposed that the low stresses observed to produce plastic deformation compared to theoretical predictions at the time could be explained in terms of the theory of dislocations.

The theory describing the elastic fields of the defects was originally developed by Vito Volterra in 1907. The term 'dislocation' referring to a defect on the atomic scale was coined by G. I. Taylor in 1934.

Until the 1930s, one of the enduring challenges of materials science was to explain plasticity in microscopic terms. A simplistic attempt to calculate the shear stress at which neighbouring atomic planes *slip* over each other in a perfect crystal suggests that, for a material with shear modulus G, shear strength τ_m is given approximately by:

$$\tau_m = \frac{G}{2\pi}.$$

As shear modulus in metals is typically within the range 20 000 to 150 000 MPa, this is difficult to reconcile with shear stresses in the range 0.5 to 10 MPa observed to produce plastic deformation in experiments.

In 1934, Egon Orowan, Michael Polanyi and G. I. Taylor, almost simultaneously realized that plastic deformation could be explained in terms of the theory of dislocations. Dislocations can move if the atoms from one of the surrounding planes break their bonds and rebond with the atoms at the terminating edge. In effect, a half plane of atoms is moved in response to shear stress by breaking and reforming a line of bonds, one (or a few) at a time. The energy required to break a single bond is far less than that required to break all the bonds on an entire plane of atoms at once. Even this simple model of the force required to move a dislocation shows that plasticity is possible at much lower stresses than in a perfect crystal. In many materials, particularly ductile materials, dislocations are the "carrier" of plastic deformation, and the energy required to move them is less than the energy required to fracture the material.

Mechanisms

A dislocation is a linear crystallographic defect or irregularity within a crystal structure which contains an abrupt change in the arrangement of atoms. The crystalline order is restored on either side of a dislocation but the atoms on one side have moved or slipped. Dislocations define the boundary between slipped and unslipped regions of material and cannot end within a lattice and must either extend to a free edge or form a loop within the crystal. A dislocation can be characterised by the distance and direction of movement it causes to atoms in the lattice and is called the Burgers vector. The Burgers vector of the dislocation remains constant even though the shape of dislocation may change.

A variety of dislocation types exist, with mobile dislocations known as glissile and immobile dislocations called sessile. The movement of mobile dislocations allow atoms to slide over each other at low stress levels and is known as glide or slip. The movement of dislocations may be enhanced or hindered by the presence other elements within the crystal and that may diffuse to the dislocation forming a Cottrell atmosphere. The interaction of hydrogen with dislocations is one of the mechanisms proposed to explain hydrogen embrittlement.

Dislocations behave as though they are a distinct entity within a crystalline material where some types of dislocation can move through the material bending, flexing and changing shape and interacting with other dislocations and features within the crystal. Dislocations are generated by deforming a crystalline material such as metals, which can cause them to initiate from surfaces, particularly at stress concentrations or within the material at defects and grain boundaries. The number and arrangement of dislocations give rise to many of the properties of metals such as ductility, hardness and yield strength. Heat treatment, alloy content and cold working can change the number and arrangement of the dislocation population and how they move and interact in order to create useful properties.

Generating Dislocations

When metals are subjected to cold working (deformation at temperatures which are relatively low as compared to the material's absolute melting temperature, T_m i.e., typically less than $0.4T_m$) the dislocation density increases due to the formation of new dislocations. The consequent increasing overlap between the strain fields of adjacent dislocations gradually increases the resistance to further dislocation motion. This causes a hardening of the metal as deformation progresses. This effect is known as strain hardening or work hardening.

Dislocation density ρ in a material can be increased by plastic deformation by the following relationship:

$$\tau \propto \sqrt{\rho}.$$

Since the dislocation density increases with plastic deformation, a mechanism for the creation of dislocations must be activated in the material. Three mechanisms for dislocation formation are homogeneous nucleation, grain boundary initiation, and interfaces between the lattice and the surface, precipitates, dispersed phases, or reinforcing fibers.

Homogeneous Nucleation

The creation of a dislocation by *homogeneous nucleation* is a result of the rupture of the atomic bonds along a line in the lattice. A plane in the lattice is sheared, resulting in 2 oppositely faced half planes or dislocations. These dislocations move away from each other through the lattice. Since homogeneous nucleation forms dislocations from perfect crystals and requires the simultaneous breaking of many bonds, the energy required for homogeneous nucleation is high. For instance, the stress required for homogeneous nucleation in copper has been shown to be $\frac{\tau_{\text{hom}}}{G} = 7.4 \times 10^{-2}$, where G is the shear modulus of copper (46 GPa). Solving for τ_{hom}, we see that the required stress is 3.4 GPa, which is very close to the theoretical strength of the crystal. Therefore, in conventional deformation homogeneous nucleation requires a concentrated stress, and is very unlikely. Grain boundary initiation and interface interaction are more common sources of dislocations.

Irregularities at the grain boundaries in materials can produce dislocations which propagate into the grain. The steps and ledges at the grain boundary are an important source of dislocations in the early stages of plastic deformation.

Frank-Read source

The Frank-Read source is a mechanism that is able to produce a stream of dislocations from a pinned segment of a dislocation. Stress bows the dislocation segment, expanding until it creates a dislocation loop that breaks free from the source.

Surfaces

The surface of a crystal can produce dislocations in the crystal. Due to the small steps on the surface of most crystals, stress in some regions on the surface is much larger than the average stress in the lattice. This stress leads to dislocations. The dislocations are then propagated into the lattice in the same manner as in grain boundary initiation. In single crystals, the majority of dislocations are formed at the surface. The dislocation density 200 micrometres into the surface of a material has been shown to be six times higher than the density in the bulk. However, in polycrystalline materials the surface sources do not have a major effect because most grains are not in contact with the surface.

Interfaces

The interface between a metal and an oxide can greatly increase the number of dislocations created. The oxide layer puts the surface of the metal in tension because the oxygen atoms squeeze into the lattice, and the oxygen atoms are under compression. This greatly increases the stress on the surface of the metal and consequently the amount of dislocations formed at the surface. The increased amount of stress on the surface steps results in an increase in dislocations.

Interaction and Arrangement

Geometrically Necessary Dislocations

Geometrically necessary dislocations are arrangements of dislocations that can accommodate a limited degree of plastic bending in a crystalline material. Tangles of dislocations are found at the early stage of deformation and appear as non well-defined boundaries; the process of dynamic recovery leads eventually to the formation of a cellular structure containing boundaries with misorientation lower than 15° (low angle grain boundaries).

Pinning

Adding pinning points that inhibit the motion of dislocations, such as alloying elements, can introduce stress fields that ultimately strengthen the material by requiring a higher applied stress to overcome the pinning stress and continue dislocation motion.

The effects of strain hardening by accumulation of dislocations and the grain structure formed at high strain can be removed by appropriate heat treatment (annealing) which promotes the recovery and subsequent recrystallization of the material.

The combined processing techniques of work hardening and annealing allow for control over dislocation density, the degree of dislocation entanglement, and ultimately the yield strength of the material.

Persistent Slip Bands

Repeated cycling of a material can lead to the generation and bunching of dislocations surrounded by regions that are relatively dislocation free. This pattern forms a ladder like structure known as a *persistent slip band* (PSB). PSB's are so-called, because they leave marks on the surface of metals that even when removed by polishing, return at the same place with continued cycling.

PSB walls are predominately made up of edge dislocations. In between the walls, plasticity is transmitted by screw dislocations.

Where PSB's meet the surface, extrusions and intrusions form, which under repeated cyclic loading, can lead to the initiation of a fatigue crack.

Geometry

Two main types of mobile dislocations exist: edge and screw. Dislocations found in real materials are typically *mixed*, meaning that they have characteristics of both.

Edge

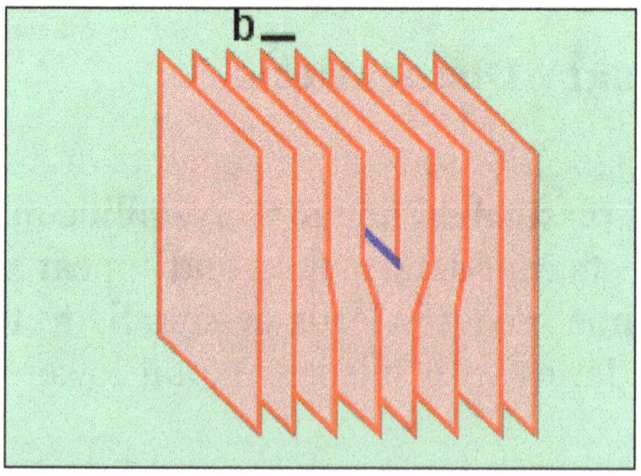

Schematic diagram (lattice planes) showing an edge dislocation. Burgers vector in black, dislocation line in blue.

A crystalline material consists of a regular array of atoms, arranged into lattice planes. An edge dislocation is a defect where an extra half-plane of atoms is introduced midway through the crystal, distorting nearby planes of atoms. When enough force is applied from one side of the crystal structure, this extra plane passes through planes of atoms breaking and joining bonds with them until it reaches the grain boundary. The dislocation has two properties, a line direction, which is the direction running along the bottom of the extra half plane, and the Burgers vector which describes the magnitude and direction of distortion to the lattice. In an edge dislocation, the Burgers vector is perpendicular to the line direction.

The stresses caused by an edge dislocation are complex due to its inherent asymmetry. These stresses are described by three equations:

$$\sigma_{xx} = \frac{-\mu \mathbf{b}}{2\pi(1-\nu)} \frac{y(3x^2+y^2)}{(x^2+y^2)^2}$$

$$\sigma_{yy} = \frac{\mu \mathbf{b}}{2\pi(1-\nu)} \frac{y(x^2-y^2)}{(x^2+y^2)^2}$$

$$\tau_{xy} = \frac{\mu \mathbf{b}}{2\pi(1-\nu)} \frac{x(x^2-y^2)}{(x^2+y^2)^2}$$

where μ is the shear modulus of the material, $\mathbf{b}$ is the Burgers vector, ν is Poisson's ratio and x and y are coordinates.

These equations suggest a vertically oriented dumbbell of stresses surrounding the dislocation, with compression experienced by the atoms near the "extra" plane, and tension experienced by those atoms near the "missing" plane.

Screw

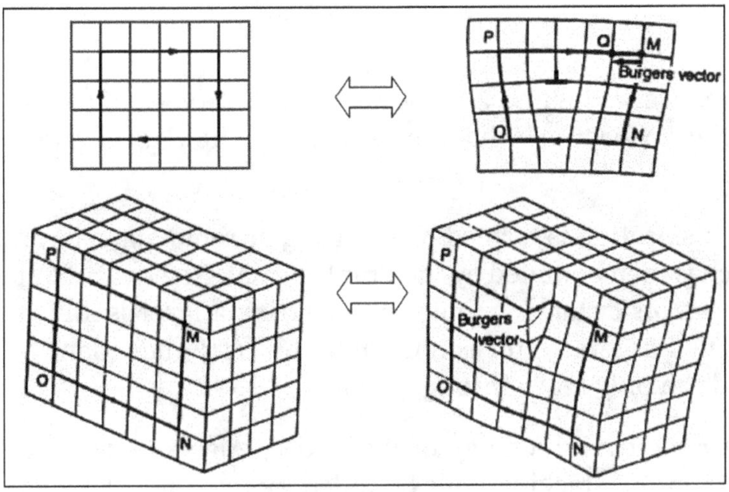

Top right: Edge dislocation.
Bottom right: Screw dislocation.

A *screw dislocation* can be visualized by cutting a crystal along a plane and slipping one half across the other by a lattice vector, the halves fitting back together without leaving a defect. If the cut only goes part way through the crystal, and then slipped, the boundary of the cut is a screw dislocation. It comprises a structure in which a helical path is traced around the linear defect (dislocation line) by the atomic planes in the crystal lattice. In pure screw dislocations, the Burgers vector is parallel to the line direction.

The stresses caused by a screw dislocation are less complex than those of an edge dislocation and need only one equation, as symmetry allows one radial coordinate to be used:

$$\tau_r = \frac{-\mu \mathbf{b}}{2\pi r}$$

where μ is the shear modulus of the material, $\mathbf{b}$ is the Burgers vector, and r is a radial coordinate. This equation suggests a long cylinder of stress radiating outward from the cylinder and decreasing with distance. This simple model results in an infinite value for the core of the dislocation at $r = 0$ and so it is only valid for stresses outside of the core of the dislocation. If the Burgers vector is very large, the core may actually be empty resulting in a micropipe, as commonly observed in silicon carbide.

Mixed

In many materials, dislocations are found where the line direction and Burgers vector are neither perpendicular nor parallel and these dislocations are called *mixed dislocations*, consisting of both screw and edge character. They are characterized by φ, the angle between the line direction and Burgers vector, where $\varphi = \pi / 2$ for pure edge dislocations and $\varphi = 0$ for screw dislocations.

Partial

Partial dislocations leave behind a stacking fault. Two types of partial dislocation are the *Frank partial dislocation* which is sessile and the *Shockley partial dislocation* which is glissile.

A Frank partial dislocation is formed by inserting or removing a layer of atoms on the {111} plane which is then bounded by the Frank partial. Removal of a close packed layer is known as an *intrinsic* stacking fault and inserting a layer is known as an *extrinsic* stacking fault. The Burgers vector is normal to the {111} glide plane so the dislocation cannot glide and can only move through *climb*.

In order to lower the overall energy of the lattice, edge and screw dislocations typically disassociate into a stacking fault bounded by two Shockley partial dislocations. The width of this stacking-fault region is proportional to the stacking-fault energy of the material. The combined effect is known as an *extended dislocation* and is able to glide as a unit. However, dissociated screw dislocations must recombine before they can cross slip, making it difficult for these dislocations to move around barriers. Materials with low stacking-fault energies have the greatest dislocation dissociation and are therefore more readily cold worked.

Stair-rod and the Lomer–Cottrell junction

If two glide dislocations that lie on different {111} planes split into Shockley partials and intersect, they will produce a stair-rod dislocation with a Lomer-Cottrell dislocation at its apex. It is called a *stair-rod* because it is analogous to the rod that keeps carpet in-place on a stair.

Jog

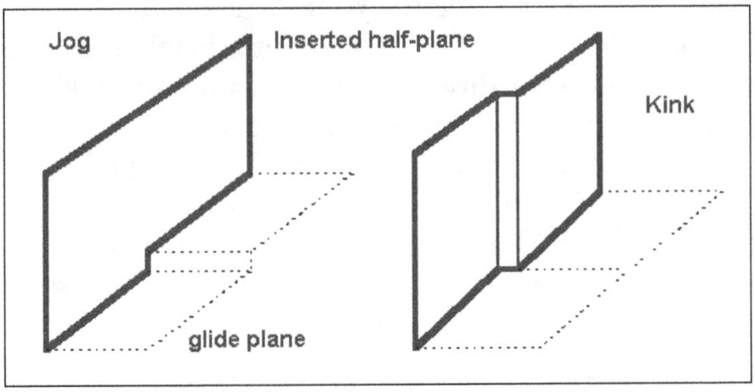

Geometrical differences between jogs and kinks.

A Jog describes the steps of a dislocation line that are not in the glide plane of a crystal structure. A dislocation line is rarely uniformly straight, often containing many curves and/or steps that can impede or facilitate dislocation movement by acting as pinpoints or nucleation points respectively. Because jogs are out of the glide plane, under shear they cannot move by glide (movement along the glide plane). They instead must rely on vacancy diffusion facilitated climb to move through the lattice. Away from the melting point of a material, vacancy diffusion is a slow process, so jogs act as immobile barriers at room temperature for most metals.

Jogs typically form when two non-parallel dislocations cross during slip. The presence of jogs in a material increases its yield strength by preventing easy glide of dislocations. A pair of immobile jogs in a dislocation will act as a Frank-Read source under shear, increasing the overall dislocation density of a material. When a material's yield strength is increased via dislocation density increase, particularly when done by mechanical work, it is called work hardening. At high temperatures, vacancy facilitated movement of jogs becomes a much faster process, diminishing their overall effectiveness in impeding dislocation movement.

Kink

Kinks are steps in a dislocation line parallel to glide planes. Unlike jogs, they facilitate glide by acting as a nucleation point for dislocation movement. The lateral spreading of a kink from the nucleation point allows for forward propagation of the dislocation while only moving a few atoms at a time, reducing the overall energy barrier to slip.

Example in two Dimensions (2D)

In two dimensions (2D) only the edge dislocations exist, which play a central role in melting of 2D crystals, but not the screw dislocation. Those dislocations are topological point defects which implies that they cannot be created isolated by an affine transformation without cutting the hexagonal crystal up to infinity (or at least up to its boarder). They can only be created in pairs with antiparallel Burgers vector. If a lot of dislocations are e. g. thermally excited, the discrete translational order of the crystal is destroyed. Simultaneously, the shear modulus and the Young's modulus disappear, which implies that the crystal is molten to a fluid phase. The orientational order is not yet destroyed (as indicated by lattice lines in one direction) and one finds - very similar to liquid crystals - a fluid phase with typically a six-folded director field. This so called hexatic phase still has an orientational stiffness. The isotropic fluid phase appears, if the dislocations dissociate into isolated five-folded and seven-folded disclinations. This two step melting is described within the so called Kosterlitz-Thouless-Halperin-Nelson-Young-theory (KTHNY-theory), based on two transitions of Kosterlitz-Thouless-type.

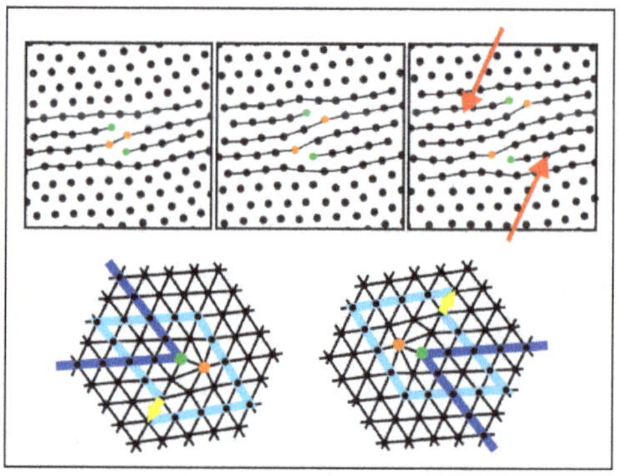

Dissociation of a pair of dislocations due to shearing (red arrows) of an hexagonal crystal in 2D. A dislocation in 2D consists of a bound pair of five-folded (green) and seven-folded (orange) coordination number.

Observation of Transmission Electron Microscopy (TEM)

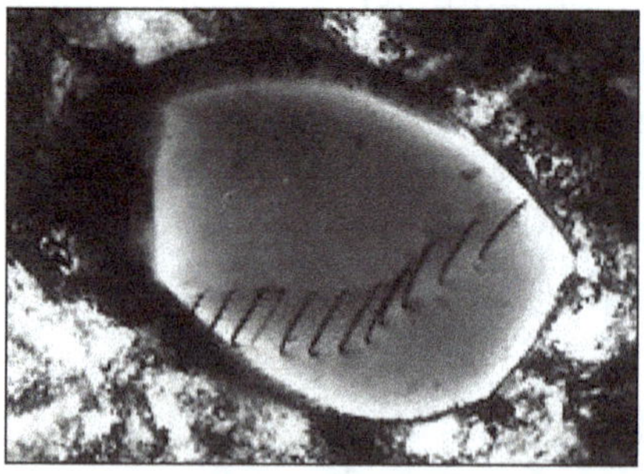

Transmission electron micrograph of dislocations.

Transmission electron microscopy can be used to observe dislocations within the microstructure of the material. Thin foils of material are prepared to render them transparent to the electron beam of the microscope. The electron beam undergoes diffraction by the regular crystal lattice planes into a diffraction pattern and contrast is generated in the image by this diffraction (as well as by thickness variations, varying strain, and other mechanisms). Dislocations have different local atomic structure and produce a strain field, and therefore will cause the electrons in the microscope to scatter in different ways. Characteristic 'wiggly' contrast of the dislocation lines as they pass through the thickness of the material in the figure (also dislocations cannot end in a crystal, and these dislocations are terminating at the surfaces since the image is a 2D projection).

Dislocations do not have random structures, the local atomic structure of a dislocation is determined by the Burgers vector. One very useful application of the TEM in dislocation imaging is the ability to experimentally determine the Burgers vector. Determination of the Burgers vector is achieved by what is known as $\vec{g}\cdot\vec{b}$ ("g dot b") analysis. When performing dark field microscopy with the TEM, a diffracted spot is selected to form the image, lattice planes diffract the beam into spots), and the image is formed using only electrons that were diffracted by the plane responsible for that diffraction spot. The vector in the diffraction pattern from the transmitted spot to the diffracted spot is the $\vec{g}$ vector. The contrast of a dislocation is scaled by a factor of the dot product of this vector and the Burgers vector ($\vec{g}\cdot\vec{b}$). As a result, if the Burgers vector and $\vec{g}$ vector are perpendicular, there will be no signal from the dislocation and the dislocation will not appear at all in the image. Therefore, by examining different dark field images formed from spots with different g vectors, the Burgers vector can be determined.

Other Methods

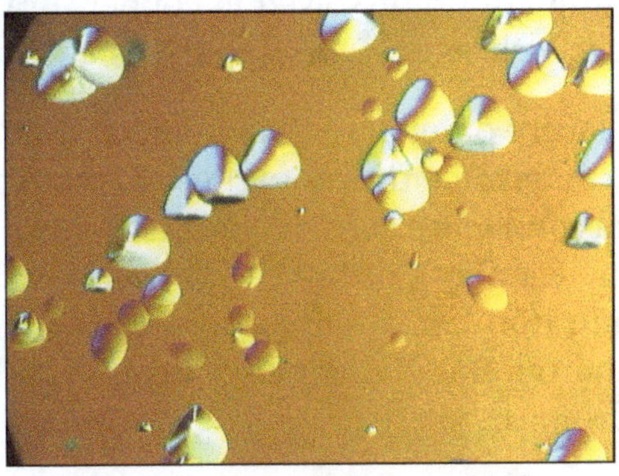

Etch Pits formed on the ends of dislocations in silicon, orientation (111).

Field ion microscopy and atom probe techniques offer methods of producing much higher magnifications (typically 3 million times and above) and permit the observation of dislocations at an atomic level. Where surface relief can be resolved to the level of an atomic step, screw dislocations appear as distinctive spiral features – thus revealing an important mechanism of crystal growth: where there is a surface step, atoms can more easily add to the crystal, and the surface step associated with a screw dislocation is never destroyed no matter how many atoms are added to it.

Chemical Etching

When a dislocation line intersects the surface of a metallic material, the associated strain field locally increases the relative susceptibility of the material to acid etching and an etch pit of regular geometrical format results. In this way, dislocations in silicon, for example, can be observed *indirectly* using an interference microscope. Crystal orientation can be determined by the shape of the etch pits associated with the dislocations.

Dislocations are one-dimensional defects caused by holes that are not large enough to be a vacancy.

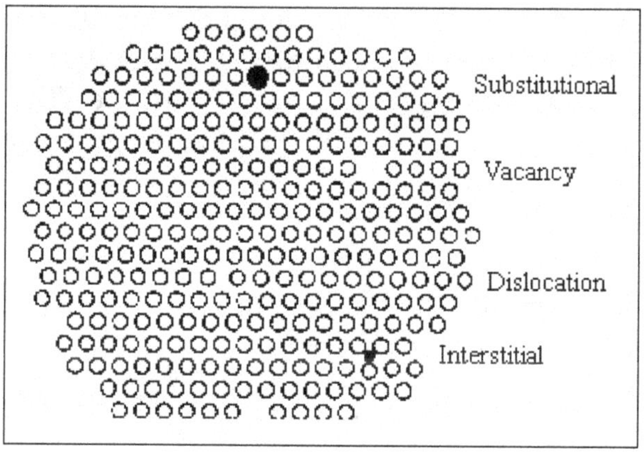

When a significant fraction of the original particles are replaced by impurities, it is possible to get a solid solution. Alloys, such as bronze and brass, are examples of solid solutions. Bronze is a solution of tin dissolved in copper. Brass is a mixture of copper and zinc that can contain as little as 10%, or as much as 45%, zinc.

Distortions of the crystal lattice often occur when impurities are added to a solid. As a result, point defects often determine the properties of a material. They can change the ease with which a material conducts electricity, its mechanical strength, its ability to be shaped by hammering (malleability), or to be drawn into wires (ductility). Dissolving small amounts of carbon in iron, for example, give the alloy known as steel, which is significantly stronger than iron. But higher percentages of carbon make steel so brittle that it can shatter when dropped.

Point defects distort the lattice and provide a way for atoms to move about the solid. Atoms can move from a lattice site into a vacancy, for example, creating a new vacancy, as shown in the figure below.

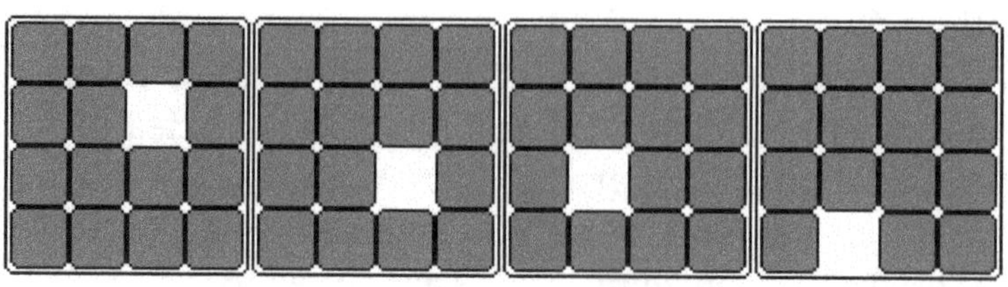

Theoretical calculations of the ease with which one plane of atoms should slip over another suggest that metals should be much more resistant to stress than they are. In other words, metals are softer than one would expect. Metallurgists have explained this by assuming that metals contain defects that allow planes of atoms to slip past each other more readily than expected. This hypothesis has been confirmed by microscopic analysis, which shows dislocations that run through the crystal. There are two types of dislocations: Edge or screw dislocations. An edge dislocation is an extra half plane of atoms that goes part way through a solid structure, as shown in the figure below.

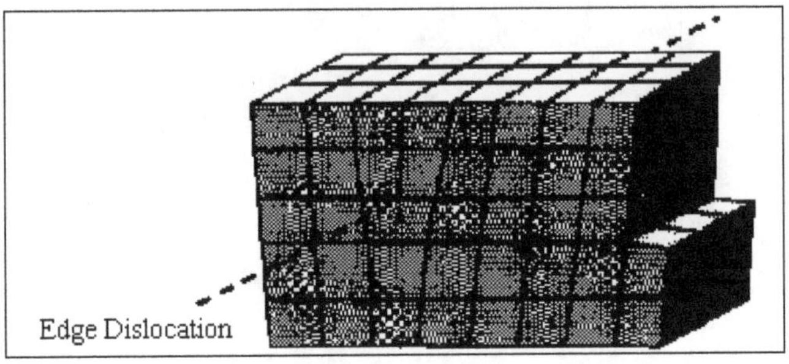

Imagine, for example, a single playing card inserted halfway into a deck of cards. The line formed by the inserted card would be a dislocation line. The presence of a dislocation defect allows one plane of atoms to slip more easily over its neighboring plane of atoms, as shown in the figure below. Not all the atoms in the two planes move past each other simultaneously; they move one row at a time.

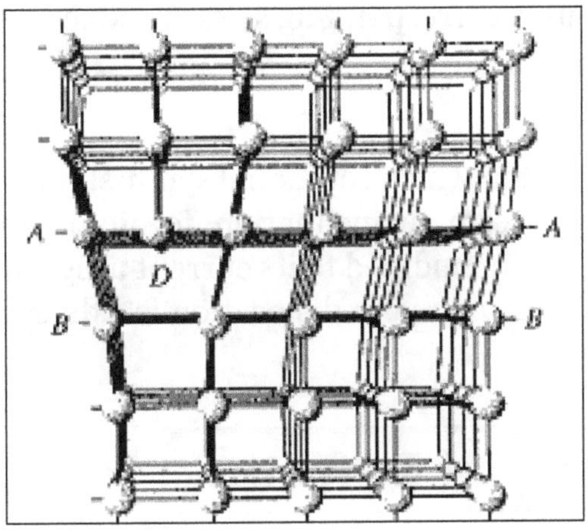

An often quoted analogy is that of moving a carpet. Dragging the carpet across the floor is difficult because of the friction developed from the contact of the surface of the carpet with the floor. Imagine what would happen, however, if a wrinkle is put into the carpet, as shown in part (a) of the figure below. The carpet can now be moved by pushing the wrinkle across the floor, because only the friction between a small section of carpet and the floor has to be overcome. A similar phenomenon occurs when one plane of atoms moves past another by means of a dislocation defect.

Because they allow planes of atoms in a solid to move one row at a time, dislocations can weaken a metal. Paradoxically, they can also strengthen a metal when the dislocations intersect to product knots similar to the intersecting wrinkles in the figure labeled "b" in the figure below. This phenomenon is encountered with metals that have been work hardened. Consider what happens, for example, when a piece of iron is heated, hammered, cooled, reheated, and reworked to form wrought iron. In the course of work hardening the metal, intersecting dislocations are generated that hinder the movement of planes of atoms.

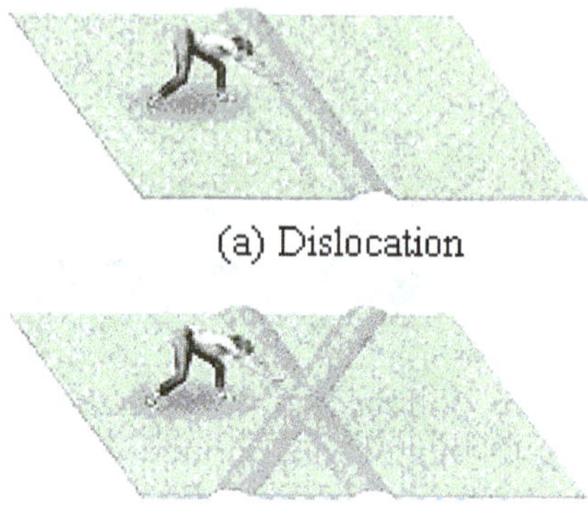

(a) Dislocation

(b) Work hardening

Screw dislocations are more difficult to visualize than edge dislocations. The figure below shows how a screw dislocation is produced when one side of a crystal is displaced relative to the other side. For either edge or screw dislocations a distortion is produced around the dislocation with a corresponding stress produced within the material.

Disclination

A disclination is a line defect in which rotational symmetry is violated. In analogy with dislocations in crystals, the term, disinclination, for liquid crystals first used by F. C. Frank and since then has been modified to its current usage, disclination. It is a defect in the orientation of director whereas a dislocation is a defect in positional order.

Two-Dimensional Defects

The two-dimensional defects that appear in crystals can be usefully divided into three types: free surfaces, which are the external surfaces at which the solid terminates at a vapor or liquid, intercrystalline boundaries, which separate grains or distinct phases within the solid, and internal defects that disrupt the crystalline pattern over a surface within a crystal. All of these defects have two important characteristics. First, since they are surfaces in a crystal they have particular atomic structures that depend on orientation. Second, they have a positive energy. The energy per unit area is ordinarily equal to the surface tension, σ, of the interface.

Free Surfaces

The interface between a crystalline solid and a vapor or liquid governs the interaction between the two phases, and influences the behavior of the solid in many important ways. It also affects the shape of the solid. The shape and properties of the free surface are determined by a combination of its structure and its energy. These two parameters are related to one another.

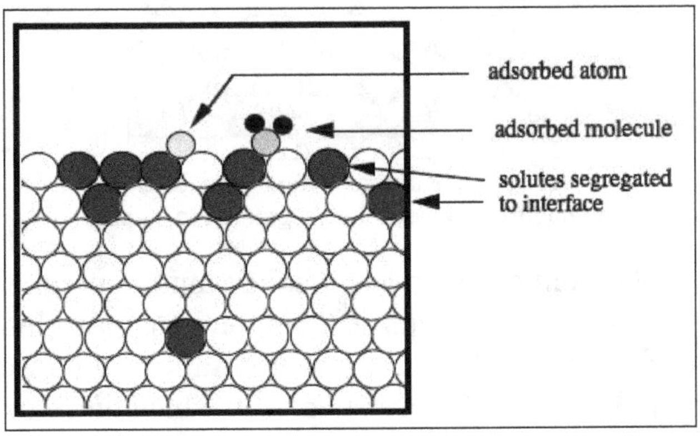

Solid surface showing segregated solutes and adsorbates.

The crystal structure of the solid extends up to the free surface. The structure of the surface is derived from the pattern of atomic packing on the crystal plane that lies parallel to the surface, and the behavior of the surface can often be inferred from that pattern. However, the structure of the surface always differs at least a bit from that of a crystal plane that parallels it within the bulk crystal. Even when the crystal sits in a perfect vacuum, the atoms in the free surface are bonded on only one side and the spacing and the configuration of the atoms in the first few planes at the interface adjusts to accommodate the asymmetry of the bonding in the bast possible way. The structure of a real interface is further complicated by bonding interactions across the interface and by chemical changes in the interfacial plane. The dangling bonds at the interface provide favorable sites for the adsorption of atoms that do not fit well into the bulk lattice. The interface is often enriched in solute species (surfactants). Oxygen, sulfur and phosphorous are common surfactants in engineering solids.

The interfacial tension of the free surface depends on its structure, and is determined, in large part, by the number of bonds that are broken because the crystal is not continued on the far side of the interface. As a consequence the interfacial tension, σ, is a function of the orientation of the surface. In a typical metal, the interfacial tension is a relatively weak function of the orientation. Because metallic bonding is relatively nondirectional, the interfacial tension of a typical metal tends to vary with the density of packing in the interface and is minimized for surface orientations that parallel closepacked planes ({111} in FCC, {110} in BCC). Covalent solids have fewer neighbors, and more directional bonds, and are much more strongly anisotropic in their surface tension.

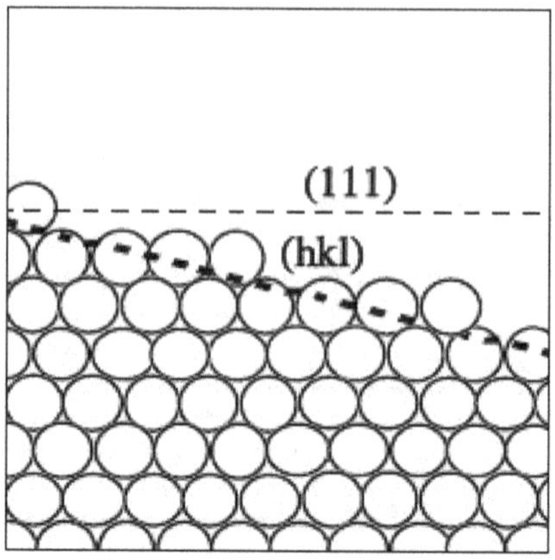

A surface of overall orientation (hkl)
made up of closepacked segments in steps.

The external surfaces of metallic and covalent solids tend to parallel the planes of closest atomic packing, as illustrated in figure. This tendency also influences the atomic structures of interfacial planes that have other orientations. An interface that does not parallel a close-packed plane often has a structure in which flat, close-packed segments form ledges of atomic height that give a stepped structure with the proper overall orientation, as illustrated in figure. While the stepped morphology increases the surface area, the overall surface tension is lower because of the relatively low surface tension of the close-packed steps.

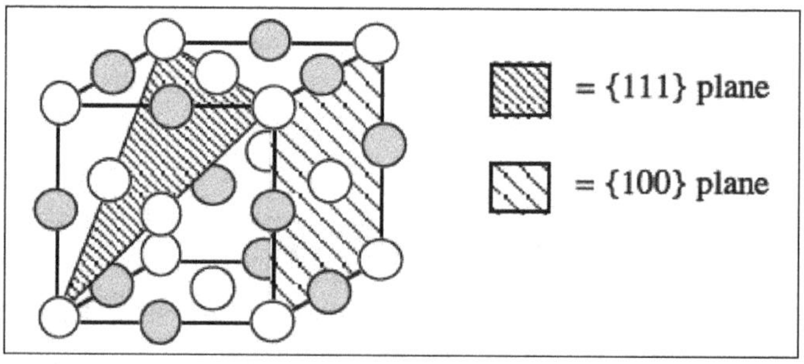

The NaCl structure, showing the electrical neutrality of the {100} plane
and the preferential charge on the {111} plane.

The interfacial tension of a surface in an ionic solid is strongly affected by the distribution of electric charge. In strongly ionic materials (for example, NaCl) the electrically neutral surfaces ({100} in NaCl) have much lower interfacial tensions than the charged surfaces (such as {111}). The relative interfacial tensions of charged and uncharged surfaces can, of course, be altered by the addition of surfactants, which are often preferentially attracted to the charged surfaces.

The Shapes of Crystals

The preferred shape of a crystal that is bounded by free surfaces is the one that minimizes its total interfacial energy. The macroscopic surface of the crystal can always be treated as if it were made up of facets, the kth one of which has area, A_k, and interfacial tension, σk. The energy is, then,

$$E = \sum_{k} \sigma_k A_k$$

When the crystal is a simple metal the surface tension is nearly isotropic. The preferred shape is nearly spherical, since a spherical shape minimizes the surface area for given volume. When the crystal is ionic or strongly covalent there are crystal planes whose tensions are, relatively, very low. To minimize the total interfacial energy, ionic and covalent solids tend to assume polygonal shapes with large, flat facets parallel to the lowenergy planes.

This is the physical reason that compound solids and minerals often look crystalline, while metals usually do not. The natural macroscopic shapes of such compounds are polygonal (crystalline) because of the strong anisotropy of the surface tension. You can readily verify, for example, that the grains of common table salt have a cubic shape; this shape reflects the low energy of the neutral {100} surfaces in the NaCl structure. Many common minerals have hexagonal structures in which the prismatic planes (the vertical planes in the hexagonal unit cell) are electrically neutral. These materials tend to grow as well-shaped hexagonal prisms. Simple metals, on the other hand, are more isotropic in their interfacial tensions, and tend to form crystals that are cylindrical or spherical in their macroscopic shapes.

References

- Crystal-defect, science: britannica.com, Retrieved 18 March, 2020
- Non-stoichiometric-defects: amazingsolids.wordpress.com, Retrieved 10 July, 2020
- Hong, J.; Hu, Z.; Probert, M.; Li, K.; Lv, D.; Yang, X.; Gu, L.; Mao, N.; Feng, Q.; Xie, L.; Zhang, J.; Wu, D.; Zhang, Z.; Jin, C.; Ji, W.; Zhang, X.; Yuan, J.; Zhang, Z. (2015). "Exploring atomic defects in molybdenum disulphide monolayers". Nature Communications. 6: 6293. Bibcode:2015NatCo...6E6293H. doi:10.1038/ncomms7293. PMC 4346634. PMID 25695374
- Kittel, Charles (2005). Introduction to Solid State Physics (8th ed.). Wiley. ISBN 978-0-471-41526-8
- Derlet, P. M.; D. Nguyen-Manh; S. L. Dudarev (2007). "Multiscale modeling of crowdion and vacancy defects in body-centered-cubic transition metals". Phys. Rev. B. 76 (5): 054107. Bibcode:2007PhRvB..76e4107D. doi:10.1103/physrevb.76.054107
- Brayfindley, Evangelina; Irace, Erica E.; Castro, Claire; Karney, William L. (2015). "Stone–Wales Rearrangements in Polycyclic Aromatic Hydrocarbons: A Computational Study". J. Org. Chem. 80 (8): 3825–3831. doi:10.1021/acs.joc.5b00066. PMID 25843555
- Geng, Hua Y.; et al. (2012). "Anomalies in nonstoichiometric uranium dioxide induced by a pseudo phase transition of point defects". Phys. Rev. B. 85 (14): 144111. arXiv:1204.4607. doi:10.1103/PhysRevB.85.144111
- Spence, J.C.H.; et al. (2006). "Imaging dislocation cores – the way forward". Philosophical Magazine. 86 (29–31): 4781–4796. Bibcode:2006PMag...86.4781S. doi:10.1080/14786430600776322
- Disclination, definition: definitions.net, Retrieved 18 June, 2020

Permissions

All chapters in this book are published with permission under the Creative Commons Attribution Share Alike License or equivalent. Every chapter published in this book has been scrutinized by our experts. Their significance has been extensively debated. The topics covered herein carry significant information for a comprehensive understanding. They may even be implemented as practical applications or may be referred to as a beginning point for further studies.

We would like to thank the editorial team for lending their expertise to make the book truly unique. They have played a crucial role in the development of this book. Without their invaluable contributions this book wouldn't have been possible. They have made vital efforts to compile up to date information on the varied aspects of this subject to make this book a valuable addition to the collection of many professionals and students.

This book was conceptualized with the vision of imparting up-to-date and integrated information in this field. To ensure the same, a matchless editorial board was set up. Every individual on the board went through rigorous rounds of assessment to prove their worth. After which they invested a large part of their time researching and compiling the most relevant data for our readers.

The editorial board has been involved in producing this book since its inception. They have spent rigorous hours researching and exploring the diverse topics which have resulted in the successful publishing of this book. They have passed on their knowledge of decades through this book. To expedite this challenging task, the publisher supported the team at every step. A small team of assistant editors was also appointed to further simplify the editing procedure and attain best results for the readers.

Apart from the editorial board, the designing team has also invested a significant amount of their time in understanding the subject and creating the most relevant covers. They scrutinized every image to scout for the most suitable representation of the subject and create an appropriate cover for the book.

The publishing team has been an ardent support to the editorial, designing and production team. Their endless efforts to recruit the best for this project, has resulted in the accomplishment of this book. They are a veteran in the field of academics and their pool of knowledge is as vast as their experience in printing. Their expertise and guidance has proved useful at every step. Their uncompromising quality standards have made this book an exceptional effort. Their encouragement from time to time has been an inspiration for everyone.

The publisher and the editorial board hope that this book will prove to be a valuable piece of knowledge for students, practitioners and scholars across the globe.

Index

www.ingramcontent.com/pod-product-compliance
Lightning Source LLC
LaVergne TN
LVHW080629210926
848485LV00003B/234